Flying Creatures: the secret meaning of birds

An Illustrated Guide to Ornithology, Spiritual Symbolism, Cultural, and Mythological Significance from Around the World.

Created by:
Mayumi Nakagaki
and
Ziggy Quinete

"Flying Creatures: The Secret Meaning of Birds" is a work of fiction. The contents of this book, including but not limited to the descriptions, illustrations, and symbolism of flowers, are intended for entertainment and educational purposes only. The authors and publishers make no claim to the accuracy or completeness of the information provided and are not responsible for any errors or omissions. This book should not be used as a reference for medical or herbal use of flowers, and no part of this book should be interpreted as medical advice. The interpretation and use of the book's contents are at the reader's own risk.

First Printing Edition, 2023
ISBN - 978-65-986163-8-0

Published by Enchanted Tones
Authors: Mayumi Nakagaki and Ziggy Quinete
Illustrations: Satoshi Watanabe

"A bird does not sing because it has an answer, it sings because it has a song."

— Chinese proverb

table of contents

the meaning of flying creatures

I've always watched birds with a kind of reverence—not for what they are, but for what they suggest. A flash of color through the trees, a call in the distance, a feather drifting in the wind—these things feel like messages. Not loud, not obvious. But insistent in their quiet way, like a secret waiting to be remembered. Over time, I began to understand that birds are more than creatures of the sky—they are symbols, archetypes, and echoes of human longing. They carry the weight of mythology and the lightness of spirit. This book is my attempt to listen closely, and to share what I've heard.

Each bird in these pages was chosen not just for its beauty or behavior, but for its deeper resonance. Some, like the owl or the dove, are familiar symbols we've grown up with. Others—mythical or culturally obscure—arrived unexpectedly, whispering their meaning like old friends I didn't know I'd missed. Writing about them became a ritual. A way of sitting with stories that felt older than me, yet somehow still mine. I approached each entry not as a scientist, but as a listener, letting the symbolism and folklore take root before adding anything of my own.

I wanted this to be more than a field guide. I wanted it to feel like an altar—something to return to, slowly, whenever the world feels too heavy or too fast. Birds have a way of reminding us that perspective is everything. They are mirrors, messengers, muses. Sometimes, they're warnings. Other times, they're comforts. But always, they carry meaning in their wings—if we are still enough to notice.

Many of these symbols are drawn from ancient traditions, but I haven't been strict about borders. Myth travels like wind, and birds do too. I allowed them to cross continents and timelines, to be both fact and fiction, reality and metaphor. The scientific names are here, yes, and the details of their habitat. But also the stories. The dreams. The emotional weather they stir when they pass overhead. I believe the symbolic and the literal don't compete—they complete each other.

This book is for anyone who has ever stopped mid-step at the sound of wings, or followed a bird's flight with a sense of something unspoken stirring inside. It's for those who believe that everything—no matter how fleeting or fragile—means something. And if you don't yet believe that, maybe a few pages in, you will.

- Mayumi Nakagaki

Albatross
(Diomedeidae)

spiritual guidance

Albatrosses are primarily found in the Southern Ocean and the North Pacific. Species such as the wandering, black-browed, and royal albatross are native to sub-Antarctic islands, while the Laysan and black-footed albatrosses inhabit the North Pacific, especially around the Hawaiian archipelago and Japanese waters. These birds thrive in cold and temperate oceanic climates, nesting on remote islands that are often rugged and windswept. They feed predominantly on squid, fish, and krill, using their keen sense of smell—a rare trait in birds—to locate food from great distances. Albatrosses are also monogamous, forming long-lasting pair bonds and engaging in elaborate mating dances that can last for years before a pair finally mates.

The albatross is deeply entwined with maritime lore. Most famously, it appears in Samuel Taylor Coleridge's The Rime of the Ancient Mariner, where it represents both a good omen and a heavy burden of guilt after it is killed. In sailor superstitions, albatrosses were believed to carry the souls of dead seafarers, and harming one was considered to bring disaster. These associations have given the bird a layered symbolic profile—at once a sign of hope, endurance, and spiritual guidance, as well as a symbol of haunting consequence. Their ability to navigate uncharted waters with ease has made them enduring metaphors for freedom, the soul's journey, and the weight of personal responsibility.

Alcedo - Kingfisher
(Alcedo atthis)

omen of good fortune

The kingfisher, particularly the strikingly vibrant Alcedo atthis, carries rich symbolic meaning rooted in ancient cultures. In Greek mythology, it is tied to the tale of Alcyone and Ceyx, a tragic couple turned into birds by the gods. Alcyone, the daughter of Aeolus, god of the winds, mourned the loss of her husband Ceyx at sea so deeply that the gods took pity and transformed them both into kingfishers. This myth is the origin of the term "halcyon days," referring to a period of calm weather in winter when kingfishers were believed to nest over the sea, calming the waves. As such, the kingfisher symbolizes peace, love, calm waters, and prosperity, and has long been seen as an omen of good fortune and tranquility. In Christian art, its blue plumage has been associated with the Virgin Mary, reinforcing its ties to purity and divine protection.

In the natural world, Alcedo atthis is native to Europe, North Africa, and large parts of Asia, often inhabiting freshwater rivers, lakes, and streams with clear water and an abundance of small fish. It thrives in temperate to warm climates, especially in areas with steep riverbanks where it can dig burrows to nest. The common kingfisher is a small bird, about 6.5 inches long, with iridescent blue and orange plumage and a sharp, dagger-like bill perfect for fishing. It's a visual marvel, often seen perched quietly before diving headfirst into the water to snatch up prey with incredible precision. Though widespread, it's vulnerable to pollution and habitat destruction, which have caused population declines in certain regions. Despite its diminutive size, the kingfisher's vivid coloration and graceful hunting technique have earned it reverence and admiration throughout the ages.

Anhinga - Snakebird
(Anhinga anhinga)

movement between worlds

The anhinga is often associated with stealth, adaptability, and transformation. Its nickname, "snakebird," comes from the way it swims—with only its slender neck and pointed head above water, resembling a serpent gliding just beneath the surface. This snake-like appearance, combined with its ability to vanish silently underwater, gives it a certain mystique. In various Native American traditions, particularly among Southeastern tribes, the anhinga is seen as a creature of duality, bridging the realms of air and water, and thus representing the ability to move between worlds, to adapt to different realities, and to strike with precision when opportunity arises. It is also a symbol of patience and strategy, often waiting motionless before launching into swift action—a quiet observer until the moment calls for decisive movement.

Native to the warmer wetlands of the Americas, the anhinga can be found from the southeastern United States through Central America and into South America. It thrives in subtropical and tropical freshwater environments, including swamps, slow-moving rivers, marshes, and shallow lakes, where it hunts fish, amphibians, and aquatic invertebrates. Unlike most water birds, the anhinga lacks waterproofing oils on its feathers, which means it becomes waterlogged and must frequently perch with its wings spread wide to dry—a distinctive and almost theatrical pose. This feature, however, allows it to dive deeper and swim with less buoyancy, making it a skilled underwater predator. It nests in colonies, often alongside herons and egrets, and while it may not carry the grandeur of an eagle or the fame of a stork, the anhinga's strange grace and quiet power have earned it a subtle but enduring place in the lore of the wetlands.

Anqa
(Arabian Mythology)

harbinger of fate

The Anqa is a majestic and ancient bird in pre-Islamic Arab folklore, said to have been created with all perfections but eventually became corrupted by time and was destined to vanish from the world. Its name, “Anqa,” roughly translates to “long-necked” or “strangled,” and it is sometimes confused or associated with the Persian Simurgh or even the phoenix, due to overlapping traits. The Anqa is described as a large, elegant, and wise creature, often said to live in seclusion on Mount Qaf—a mythical mountain at the edge of the world. In many versions of the myth, it lives for 1,700 years, appearing only rarely, and its appearance is usually a harbinger of fate or divine knowledge. Symbolically, the Anqa represents wisdom, mystery, prophecy, and the unreachable—a creature too noble and too rare to belong to the mundane world.

Though not tied to any real species, the Anqa's myth served an important purpose in Arab oral tradition, especially in the centuries leading up to and just after the rise of Islam. It often functioned as a metaphor for the unknowable will of the divine or the inevitability of destiny. Unlike the phoenix, which dies and is reborn, the Anqa tends to disappear entirely, suggesting themes of loss, exile, and the fleeting nature of perfection. In Islamic-era texts, it was sometimes reinterpreted through philosophical and Sufi lenses as a spiritual ideal—something the seeker longs to reach but can never truly capture. Over time, its legend faded, replaced by other symbolic birds in Islamic poetry and philosophy, but the echo of the Anqa still lingers as a symbol of that which is both magnificent and tragically out of reach.

Anzu (Zu)
(Ancient Mesopotamia Mythology)

conjurer of storms

Anzu is a mythical bird-lion hybrid from Sumerian, Akkadian, and Babylonian myths, depicted as a giant bird with the head of a lion and sometimes with the ability to breathe fire or conjure storms. Its most famous tale appears in the Epic of Anzu, where it steals the Tablet of Destinies from the god Enlil. This theft throws the divine and earthly realms into chaos, as the tablet grants supreme authority over the universe. The story follows the gods' struggle to retrieve it, ultimately leading to Anzu's defeat by the warrior god Ninurta (in some versions, Marduk). Symbolically, Anzu represents disruption, divine rebellion, and the struggle for cosmic order, as well as the awe-inspiring, often uncontrollable power of nature. His story also raises themes of ambition, betrayal, and the limits of divine power.

The mythology of Anzu originates in ancient Mesopotamia, particularly among the Sumerians and Akkadians around 2000 BCE, and was preserved in clay tablets inscribed in cuneiform. Unlike many mythological birds that serve as protectors or guides, Anzu is an antagonist—a being who challenges the hierarchy of the gods. However, this role doesn't make him entirely evil; like many figures in Mesopotamian lore, his actions reflect the complexity of divine forces and the necessity of conflict in establishing order. Some scholars interpret Anzu as a personification of the thundercloud or the storm-wind, tying him to the volatile Mesopotamian climate and its life-sustaining yet destructive power. Over time, the figure of Anzu influenced other regional myths, possibly contributing to the imagery of later storm or chaos beasts in Middle Eastern lore.

Arctic Tern
(Sterna paradisaea)

the rhythm of life

The Arctic tern is a powerful emblem of endurance, navigation, and cyclical transformation. Known for having the longest migration of any animal on Earth, this small seabird travels from its Arctic breeding grounds to the Antarctic and back again each year—a staggering round trip of nearly 50,000 miles. Because of this incredible journey, it has come to represent not just physical stamina, but spiritual movement, global perspective, and the rhythm of life. For many coastal Indigenous cultures in the far north, the tern is seen as a seasonal signal, a messenger that marks the turning of the Earth and the return of life during the harsh Arctic thaw. Its path across hemispheres also links it symbolically to balance between extremes—light and dark, north and south, beginnings and endings.

The Arctic tern breeds along coastal and tundra regions of the Arctic, from Alaska and Canada to Greenland, Scandinavia, and Siberia. Once breeding is complete, it embarks on its epic journey to the Antarctic, making use of favorable wind currents to essentially follow the summer season around the globe. It thrives in polar and subpolar climates but also traverses temperate and tropical zones during migration. Despite its slight build—typically only about 15 inches long including tail—it is an elegant and agile flier, with deeply forked tails and a high-pitched call that pierces through the icy wind. Arctic terns feed on small fish and invertebrates, diving from the air into the sea with precision. They mate for life and are fiercely protective of their nesting grounds. Remarkably, because of their constant migration between poles, they experience more daylight than any other creature on Earth—a living symbol of light-chasing wanderers and the ever-moving cycle of time.

Avelarium
(English Folklore)

cosmic balance

The Avalerion is a mythical bird found in bestiaries and medieval lore, particularly within English and French traditions, though it may have drawn inspiration from earlier classical or Eastern myths. Described as the rarest bird in existence, it was said that only one pair lived in the world at a time. This divine couple would produce two eggs every sixty years, hatch their young, and then—mysteriously—drown themselves, leaving their offspring to carry on the cycle. Such an unusual narrative ties the Avalerion to the symbolism of renewal, purity, self-sacrifice, and cosmic balance. Like the phoenix, it echoes the themes of death giving way to life, but in a much more solemn and less fiery manner. Its name is sometimes linked linguistically to "valor" or "aile" (wing), adding associations with nobility and sacred flight.

The Avalerion doesn't belong to any known species, of course, but in medieval bestiaries—illustrated moral encyclopedias popular across Europe—it was often drawn to resemble a majestic crane, stork, or eagle, imbued with regal posture and luminous feathers. These bestiaries were not merely catalogs of animals, real or imagined—they were tools for teaching moral lessons, with each creature embodying spiritual virtues or vices. The Avalerion, by virtue of its rarity and noble self-destruction, came to represent divine uniqueness, the fleeting presence of perfection, and the idea that true legacy is left not through conquest, but through sacrifice and continuity. While the bird faded from prominence after the Middle Ages, echoes of it can be seen in later literary symbols of rare and vanishing beauty.

Avocet
(Recurvirostra avosetta)

grace and elegance

The avocet is a symbol of balance, elegance, and subtle defiance. With its slender body, striking black-and-white plumage, and uniquely upturned bill, the bird is visually arresting. Its curved beak has long been interpreted as a sign of uniqueness and nonconformity—doing things its own way, gracefully yet with precision. In some folkloric and artistic traditions, particularly in Europe, the avocet has been associated with peaceful persistence and spiritual poise, often standing as a reminder to move with intention and gentleness. Because of its refined appearance and measured movements, it also symbolizes grace under pressure—a being that navigates harsh and shifting environments with seemingly effortless calm.

Native to Europe, central Asia, and parts of Africa, the pied avocet thrives in shallow wetlands, salt pans, estuaries, and mudflats. It favors temperate to warm climates, where it can wade gracefully through water in search of small invertebrates, which it sweeps from the surface using that distinctive bill in a side-to-side motion. The avocet is also known for its protective nature; during nesting season, it can become surprisingly aggressive, diving at predators or intruders near its colony. Once endangered in parts of Europe, especially the UK, it became a symbol of conservation success after targeted efforts helped it recover—so much so that the Royal Society for the Protection of Birds (RSPB) adopted it as their emblem. Beyond its beauty, the avocet reminds us of how fragile and resilient nature can be when given a chance to thrive.

Bald Eagle
(Haliaeetus leucocephalus)

freedom and strength

The bald eagle stands as a universal emblem of freedom, vision, and strength, and is perhaps most famously known as the national bird and symbol of the United States. Chosen in 1782 to adorn the Great Seal, the bald eagle represents independence, resilience, and the soaring spirit of a nation. In Native American cultures, long before it became a national icon, the eagle held deep spiritual significance. Many tribes viewed it as a messenger between the human world and the divine, its feathers considered sacred tools in ceremonies, prayers, and rites of passage. Its sharp gaze, powerful flight, and high nesting habits connect it with the upper realms of the spirit, making it a powerful totem for clarity of purpose, courage, and elevated thought. Its "bald" appearance, in reality a brilliant white head contrasting with a dark body, enhances its regal profile and contributes to its symbolism of duality —earth and sky, life and vision.

The bald eagle is native to North America, found near large bodies of open water from Alaska and Canada all the way to northern Mexico. It thrives in temperate to subarctic climates, particularly where tall trees or cliffs provide safe nesting sites and fish are abundant. As a bird of prey, it feeds primarily on fish, but it is also an opportunistic feeder, scavenging carrion or stealing meals from other birds. Despite its status today, the bald eagle once faced near extinction in the 20th century due to hunting, habitat loss, and pesticide use —particularly DDT. Conservation efforts, including legal protection and habitat restoration, led to a remarkable recovery, and it was removed from the U.S. endangered species list in 2007. Today, it symbolizes not only freedom and pride but also the power of renewal and environmental stewardship, a living reminder of what can be reclaimed when nature is respected and protected.

Barn Owl
(Tyto alba)

spiritual insight

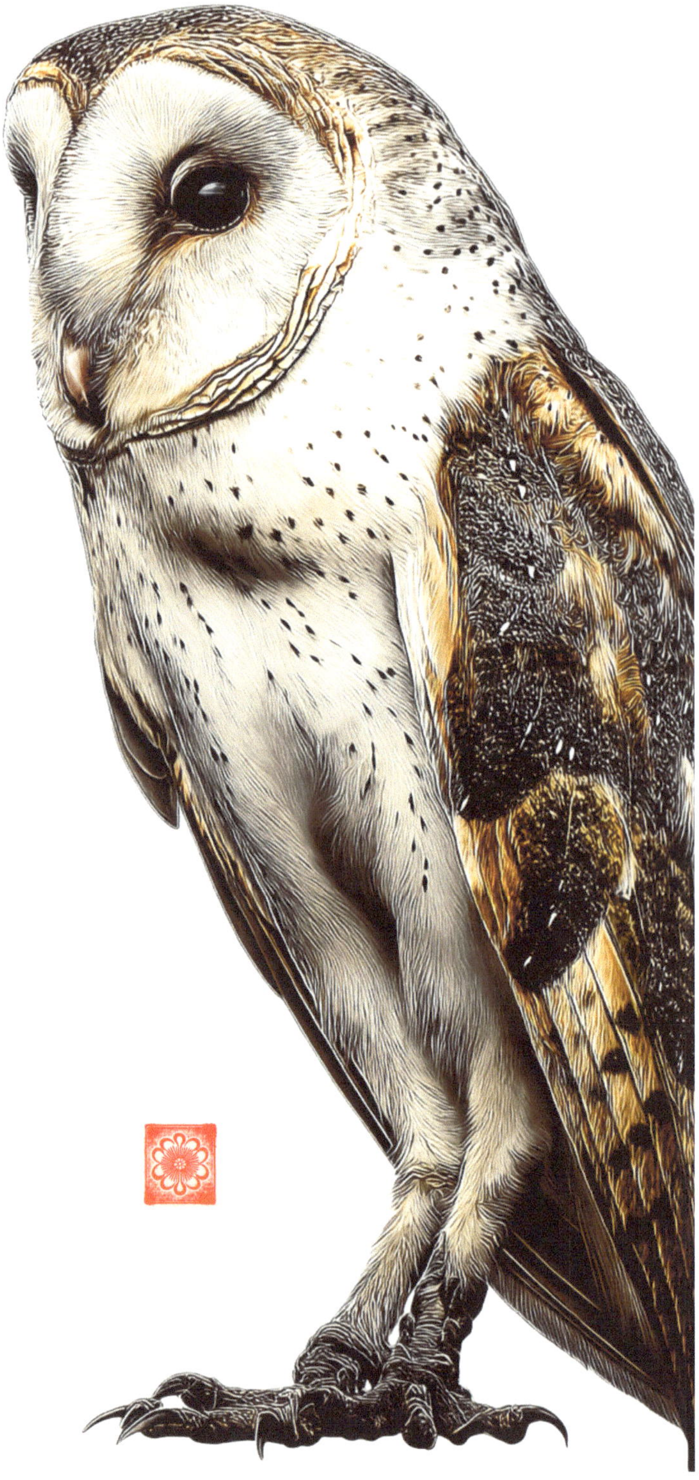

The barn owl is one of the most symbolically rich birds in human folklore, often associated with mystery, wisdom, death, and spiritual insight. Its pale, heart-shaped face, haunting screech, and silent flight have made it a subject of both reverence and fear across cultures. In Western traditions, especially in Celtic and English folklore, it has been linked to omens, the supernatural, and the spirit world. Some saw it as a messenger of death or misfortune, while others believed it could guide lost souls or reveal hidden truths. In contrast, in many Native American beliefs, the barn owl was respected for its acute senses and considered a keeper of sacred knowledge, especially knowledge of the unseen. Its presence often symbolizes a threshold moment—a crossing between worlds, or the need to confront hidden truths. Because it hunts at night and is rarely seen during the day, it also represents intuition, feminine energy, and the ability to navigate darkness.

Widely distributed across every continent except Antarctica, the barn owl is a true global citizen, thriving in a variety of environments from rural farmlands to grasslands, marshes, and even city outskirts. It prefers temperate to warm climates, especially areas with open fields for hunting and quiet structures like barns, ruins, or tree hollows for nesting. What sets the barn owl apart is its uncanny silence in flight, thanks to specialized feather structures that muffle sound, allowing it to swoop down on unsuspecting prey—usually rodents—with surgical precision. It relies heavily on its extraordinary hearing, which allows it to locate prey in total darkness. Despite its ghostly reputation, the barn owl plays a critical role in pest control and is often welcomed by farmers. Its haunting beauty and mysterious demeanor have earned it a place not only in folklore, but also in literature, films, and esoteric symbolism—as a guide through the veiled paths of the unconscious.

Barn Swallow
(Hirundo rustica)

harbinger of spring

The barn swallow is a powerful symbol of hope, renewal, loyalty, and safe return. One of the most beloved migratory birds in the world, it has long been associated with the arrival of spring and the end of hardship. In European folklore, it was believed that swallows brought good luck to a household, and harming one could bring misfortune. Sailors revered the barn swallow too, often tattooing its image on their bodies to signify a safe voyage—and one swallow tattoo for each 5,000 nautical miles traveled. The swallow became a symbol of the returning soul, of journeys made and homes re-found. In Chinese symbolism, it represents fidelity, success, and the blessing of happy marriages, while in ancient Greek myths, it was tied to the soul and to divine messengers. Universally, the swallow's yearly migration and return made it an emblem of cyclical time, perseverance, and faith in renewal.

Native to almost every continent except Antarctica, the barn swallow is incredibly adaptable and widely distributed. It breeds across North America, Europe, Asia, and North Africa, and migrates to South America, sub-Saharan Africa, and South Asia for the winter. These birds thrive in open country, farmland, and rural areas, often nesting under the eaves of buildings, bridges, and—of course—barns. They prefer temperate climates during breeding and warmer, subtropical zones for wintering. Recognizable by their deeply forked tails, glossy blue backs, and rusty-colored throats, barn swallows are agile aerial acrobats, catching insects mid-flight with elegance. They are social, often nesting in loose colonies, and fiercely devoted to their mates and nesting sites. Beyond their beauty, they serve as a reminder of nature's rhythms and of the enduring instinct to find our way home—no matter how far we roam.

Basan
(Japanese Mythology)

illusion and misdirection

The Basan, also known as the Basabasa, is a mythical fire-breathing bird found in the yokai (spirit or supernatural being) traditions of Japanese mythology, particularly in the Ehime Prefecture on the island of Shikoku. Described as resembling a large, phantom-like chicken or rooster, it is covered in shimmering white feathers and has an uncanny ability to breathe ghostly, cold fire—a flame that doesn't burn but flickers red in the darkness. The creature is said to wander mountain villages and abandoned places at night, fluttering softly with a rustling "basabasa" sound, from which its name is derived. Though it might sound fearsome, the Basan is generally considered harmless, more of a spectral curiosity than a threat. Symbolically, it's associated with the unknown, the boundaries between life and death, and the quiet magic of the in-between hours, like twilight or deep night. Its cold flame also connects it with themes of illusion, misdirection, and the supernatural's quiet intrusion into daily life.

Unlike Western mythical fire-birds that symbolize rebirth or divine glory, the Basan reflects a uniquely Japanese approach to yokai—beings that are subtle, elusive, and layered with local storytelling. The Basan is often mentioned in Edo-period encyclopedias of monsters, such as the Hyakkai Zukan by Sawaki Suushi and the Gazu Hyakki Yagyō by Toriyama Sekien. In these works, yokai like the Basan serve both to entertain and to embody regional superstitions, unexplained phenomena, or the spiritual essence of natural things. The fact that the Basan emits a flame that does not burn and appears only at night reinforces its symbolic role as a liminal creature—existing at the edge of visibility and reason. Today, it's one of the more obscure yokai, but it remains a fascinating example of how Japanese mythology often personifies the strange and wondrous qualities of the natural world in ghostly, poetic forms.

Blackbird
(Turdus merula)

hidden knowledge

The blackbird holds a powerful place in symbolic lore, representing mystery, intuition, the subconscious, and hidden knowledge. Its jet-black feathers and melodic, flute-like song have made it a recurring presence in Celtic, Christian, and alchemical traditions. In Celtic mythology, blackbirds were considered guardians of the Otherworld, their song acting as an invitation into the unseen realms. The "blackbird of the celestial world" was said to sing from the World Tree, guiding souls or luring seekers into states of dream or transformation. In Christianity, the blackbird sometimes represents temptation or the human struggle with sin, due to its dark color, yet its song is often heard as a divine call. In modern Western symbolism—especially through literature and music—it has come to represent the unknown, the creative mind, and the beauty found in melancholy. Think of the Beatles' "Blackbird," where the bird becomes a metaphor for perseverance and liberation in the face of oppression.

In the natural world, the Eurasian blackbird (Turdus merula) is native to Europe, North Africa, and much of Asia, and has also been introduced to Australia and New Zealand. It thrives in temperate climates, ranging from woodlands and gardens to urban parks. The male blackbird is known for its velvety black plumage and golden-yellow beak and eye-ring, while the female is brown and speckled, often mistaken for a different species. One of the bird's most enchanting features is its complex, lyrical song, usually delivered from a high perch at dawn and dusk. It is a territorial and monogamous species, with pairs returning to familiar nesting spots each season. Though it may appear plain at first glance, the blackbird's haunting call and secretive habits make it a living emblem of the poetic and the unseen, a symbol that whispers to those who are willing to listen in the stillness.

Blue Jay
(Cyanocitta cristata)

trickster

Symbolically, the blue jay is tied to intelligence, communication, curiosity, and assertiveness. In many Indigenous North American traditions, the blue jay is seen as a messenger and trickster, much like the raven or coyote, though with its own unique flair. Its brilliant blue plumage and loud, unmistakable call often mark its presence before it's even seen—making it a symbol of truth-telling and clarity, but also of audacity and fearlessness. The blue jay is known to mimic the calls of hawks to trick other birds and protect its territory, which adds layers of meaning: strategy, deception, and the manipulation of perception. In spiritual symbolism, it's often linked to the throat chakra—the center of expression and voice—making it a totem for those learning to speak up, assert themselves, or navigate the balance between being heard and being wise.

In terms of habitat, the blue jay is native to eastern and central North America, commonly found from southern Canada through the eastern and central United States. It thrives in deciduous and mixed forests, particularly near oak trees (due to its fondness for acorns), but it's also a regular in suburban gardens and city parks. It prefers temperate climates, though its range has slowly expanded westward. Blue jays are highly social and intelligent birds; they're part of the corvid family, which includes crows and ravens, and they display advanced problem-solving skills, tool use, and even play behaviors. They are also notorious for their bold, sometimes aggressive interactions with other birds, especially at feeders or nesting sites. Despite their loud, attention-grabbing presence, blue jays are also fiercely loyal to their mates and can be quite nurturing. With their stunning appearance and multifaceted behavior, they embody a perfect balance of brains, beauty, and boldness.

Bluebird
(Sialia sialis)

new beginnings

The bluebird is perhaps one of the most universally recognized symbols of joy, good fortune, and contentment. In Native American lore, especially among the Iroquois and Navajo, the bluebird's morning song heralds the rising sun and new beginnings, often serving as a seasonal indicator that spring has arrived. In European and American folklore, the bluebird represents domestic happiness, modesty, and inner peace, appearing in poems, fairy tales, and spiritual songs. It reached iconic status in the early 20th century through the popular song "Bluebird of Happiness," which cemented its role as a harbinger of emotional fulfillment and well-being. The bluebird is often seen as a messenger of calm and cheer, offering a sense of comfort to those facing sadness or doubt. As a spirit animal or totem, it encourages people to embrace simplicity, speak gently, and reconnect with the joy of living in the present moment.

Biologically, the eastern bluebird (Sialia sialis) is native to North America, ranging from southern Canada through the eastern United States and into parts of Mexico and Central America. It thrives in open woodlands, meadows, orchards, and suburban areas, preferring temperate climates with accessible nesting cavities. Bluebirds often nest in tree holes or man-made nest boxes and are known for their vibrant blue plumage, reddish chests, and soft, warbling songs. They feed primarily on insects and berries, often perching patiently before darting to catch their prey. Bluebirds are monogamous and highly attentive parents, often raising multiple broods in a season. Once in decline due to habitat loss and competition from invasive species like starlings, bluebird populations have made a notable recovery thanks to conservation efforts and widespread community nesting programs. More than just a visual delight, the bluebird's presence is a living embodiment of hope restored and harmony gently regained.

Brambling
(Fringilla montifringilla)

soul's journey

Symbolically, the brambling represents resilience, community, and the quiet magic of winter's arrival. Unlike louder or more flamboyant birds, the brambling is modest in both demeanor and mythological presence, yet in many Northern European traditions, its appearance was a subtle marker of change—the ushering in of cold weather and the cyclical rhythm of the seasons. In folklore, it was often associated with preparation and survival, a reminder that times of stillness and retreat are natural and necessary. Because bramblings often gather in large flocks during migration or wintering, they also carry a message of communal strength and unity in times of scarcity. Their transient nature symbolizes the soul's journey, and their return in spring is seen as a quiet promise of hope and renewal after hardship.

Native to the boreal forests of Scandinavia, northern Russia, and parts of Siberia, the brambling breeds in cool, temperate to subarctic climates, preferring dense coniferous or mixed woodlands. In winter, it migrates to more temperate parts of Europe, northern Africa, and parts of Asia, sometimes appearing in enormous flocks numbering in the hundreds of thousands, especially in areas rich in beech mast. Males are easily identified during breeding season by their bold orange breasts, black heads, and white rumps, while females are slightly more muted in coloration. Unlike many songbirds, the brambling is less territorial, often nesting in proximity to others, which reflects its symbolic tie to peaceful coexistence and mutual support. While it may not be a cultural icon like the bluebird or raven, the brambling's quiet presence and adaptive spirit make it a beautiful symbol of weathering storms with grace and trusting the return of light after the dark.

Bullfinch
(Pyrrhula pyrrhula)

messenger from the spirit world

The bullfinch carries symbolic themes of melancholy, remembrance, inner strength, and quiet devotion. In European folklore, particularly in the British Isles and Germanic regions, the bullfinch was often viewed as an omen of death or a messenger from the spirit world, not because of anything sinister, but due to its soft, mournful call and elusive presence in the winter landscape. In older rural superstitions, its appearance near the home—especially in pairs—was thought to signify an impending loss or transition. Yet this association wasn't one of dread, but of solemn acceptance, making the bullfinch a symbol of grief tempered by grace, of transformation through stillness. In more modern interpretations, especially in literature and poetry, the bullfinch is also tied to modesty, beauty in restraint, and the kind of emotional depth that doesn't need to shout to be felt.

Found across Europe and into parts of temperate Asia, the Eurasian bullfinch thrives in coniferous forests, woodland edges, hedgerows, and quiet gardens, particularly in cooler climates. It prefers temperate to subarctic regions, and is especially associated with snowy winter scenes, where its vivid coloring makes a poetic contrast against the frost. The male is especially eye-catching, with a rosy-red breast, soft grey back, and a distinct black cap and face, while the female wears more muted tones. Despite its beauty, the bullfinch is famously shy and tends to move in small, quiet groups. It feeds on seeds, buds, and berries—sometimes to the frustration of orchard-keepers, as it has a particular taste for fruit tree blossoms. Though not a prolific singer, its low, flute-like whistle carries a tender, introspective quality that aligns beautifully with its symbolism. In its simplicity, the bullfinch reminds us that not all power is loud—some is found in silence, in presence, and in the beauty of gentle resilience.

Buzzard
(Buteo buteo)

renewal

The buzzard is linked to purification, transformation, and watchful awareness. In many cultures, particularly across Celtic and Indigenous American traditions, birds of prey that scavenge—like buzzards or vultures—are not feared, but revered as sacred agents of nature's balance. The buzzard, with its patient flight and slow, methodical movements, represents the ability to see the bigger picture, to wait for the right moment, and to transform decay into sustenance. It embodies the idea that death is not an end, but part of a natural cycle of rebirth and nourishment. In spiritual symbolism, the buzzard is often seen as a guide through difficult transitions, one that teaches how to let go, cleanse spiritually, and rise with new perspective. Its presence is both grounding and elevating, reminding us that even what seems undesirable has a sacred purpose in the grand ecosystem of life.

The buzzard is native to Europe, parts of North Africa, and western Asia, and is especially common in the British countryside, where it has made a significant comeback after being nearly wiped out in some regions during the 19th and 20th centuries. It thrives in temperate climates, particularly in open landscapes such as farmland, moorland, and forest edges, where it can soar effortlessly on thermals and scan the ground for prey. Despite being a bird of prey, the buzzard is opportunistic and will scavenge when necessary, feeding on carrion, small mammals, and birds. It's often seen circling in the sky with broad, rounded wings and a fanned tail, releasing a distinctive mewing call that echoes through the air. Its slow, deliberate nature and its ability to adapt to both wild and semi-urban environments make it a survivor—a quiet guardian of the threshold between life and death. And while it may not command the awe of an eagle, the buzzard offers something more subtle: the wisdom of patience and the beauty of nature's necessary recycling.

Caladrius
(Roman Mythology)

healing

The Caladrius is a mythical white bird, often depicted as dove-like or stork-like, that resides in royal courts or holy temples. According to Roman and later medieval belief, the Caladrius possessed an extraordinary gift: it could diagnose illness and absorb disease. If a sick person looked into its eyes and the Caladrius turned its gaze away, the person was destined to die. But if it locked eyes and looked upon them with compassion, the disease would pass from the person into the bird, which would then fly into the sky, dispersing the sickness into the sun and purifying itself and the patient. Because of this, the Caladrius came to symbolize not just healing, but sacrificial purity, divine grace, and the channel between the physical and spiritual realms. In Christian bestiaries, it was interpreted as a Christ figure—pure, willing to take on humanity's suffering, and capable of miraculous restoration. The Caladrius was also said to be entirely white, symbolizing its freedom from sin and its celestial nature.

Though not based on any single real species, scholars believe the Caladrius may have been inspired by pale birds common in royal or religious iconography, such as white doves, egrets, or albino animals revered for their rarity. Its story flourished throughout the Middle Ages in illuminated manuscripts and sermons, where it was used as a metaphor for moral purity, divine justice, and the possibility of spiritual renewal. The Caladrius also appears in later esoteric texts as a symbol of psychic healing, spiritual empathy, and celestial wisdom, sometimes reimagined as a phoenix-like figure in alchemical thought. Though it's largely forgotten in modern lore, its legacy lingers as a representation of the mysterious, miraculous power of healing—and the idea that sometimes salvation comes not through force, but through gentle, silent mercy.

Canary
(Serinus canaria)

awakening of the soul

The canary is widely seen as a symbol of happiness, vitality, freedom, and emotional expression. Known for its bright yellow plumage and melodious song, the canary has long been associated with light-heartedness and the pure joy of being alive. In folklore and spiritual symbolism, it often represents the awakening of the soul, the power of music to heal, and the importance of letting one's true voice be heard. Interestingly, the canary also holds a more complex symbolic role through its historic use in coal mines, where miners brought canaries underground to warn them of invisible toxic gases. In this context, the bird became an emblem of sensitivity and sacrifice, a living warning system whose distress signaled danger. As such, the canary embodies a kind of emotional intelligence—the ability to sense what others can't and to give voice to the unspoken, whether in joy or distress. It's both a songbird of sunshine and a sentinel of survival.

Native to the Macaronesian Islands (specifically the Canary Islands, Azores, and Madeira), the wild canary was first discovered by Europeans in the 15th century. It thrives in mild, subtropical climates, especially in semi-open woodlands and scrubby hillsides. The species was later domesticated and bred widely throughout Europe, becoming especially popular in the 17th century as a caged songbird. Selective breeding produced a variety of vibrant colors and song types, turning the canary into a symbol of refinement and companionship in aristocratic homes and artistic salons. Today, domestic canaries are known for their sweet temperaments and intricate vocalizations—particularly in the males, who sing to attract mates and mark territory. Despite their long history of captivity, canaries still represent the call for liberation, the beauty of honest self-expression, and the joy that comes from living lightly and fully.

Cardinal
(Cardinalis cardinalis)

faith and the divine

The cardinal is deeply symbolic of life force, faith, and the presence of the divine, especially in the context of personal loss and spiritual messages. In many cultures, particularly in North American folklore and Christian symbolism, it's believed that when a cardinal appears, it's a visitation from a loved one who has passed—a messenger bridging the seen and unseen worlds. Its red color, vibrant even against the bleakest winter landscape, makes it a living symbol of endurance, passion, and the fire of the soul that persists through hardship. Spiritually, it's often associated with the heart chakra and vitality, representing the courage to live boldly and to honor love and memory without fear. In Christian art, the cardinal's red plumage has been linked to the blood of Christ and the robes of high clergy, marking it as a creature of faith, leadership, and divine authority. Its strong, clear call is also symbolic—seen as a voice that refuses to be silenced, one that encourages self-expression and standing tall in one's truth.

The northern cardinal is native to North and Central America, with its range extending from southern Canada through the eastern and central United States, into Mexico and parts of Central America. It thrives in woodlands, shrublands, gardens, and suburban areas, and is one of the few songbirds in North America where both male and female sing, though the male is especially known for his bright coloring and bold song. Cardinals prefer temperate to warm climates and are year-round residents rather than migratory, making them a familiar and comforting sight in winter. The male's vivid crimson feathers and the female's warm, dusky hues are both beautiful and symbolic of the balance of strength and subtlety, action and reflection. Mated pairs often remain together year-round and engage in a charming courtship behavior where the male feeds the female beak-to-beak. With their enduring presence, emotional symbolism, and striking appearance, cardinals remain one of nature's most potent reminders that love, spirit, and vitality are never far away—they're just outside the window, waiting to sing.

Cassowary
(Casuarius casuarius)

spirits of the forest

The cassowary represents protection, ancestral strength, and the hidden power of the land. In the traditions of Indigenous peoples of northern Australia and Papua New Guinea, the cassowary is often seen as a guardian of the forest, a sacred creature connected to the spirit world and the ancient Earth itself. Some tribes believe the cassowary to be a totem animal, embodying fierce independence, the link between humans and nature, and the deep power of femininity, as females are larger and more dominant. Its dramatic appearance—jet-black body, electric-blue neck, fiery red wattles, and a bony casque atop its head—adds to its mystical aura. The casque, in particular, is said to allow the cassowary to communicate with the spirit realm or move through time, and some cultures associate the bird with death and rebirth, due to its solitary, almost ghostly presence in the deep rainforest. It is also a symbol of territorial power—stand your ground, move with purpose, and defend what is yours.

The cassowary is native to tropical rainforests of northeastern Australia, Papua New Guinea, and surrounding islands, where it inhabits humid, dense, lowland forests and occasionally mangroves. It thrives in hot, wet, equatorial climates, living a largely solitary life hidden beneath the thick canopy. Though flightless, the cassowary is remarkably fast and powerful, capable of running up to 30 miles per hour and delivering lethal kicks with its dagger-like inner claws—earning it the title of "the world's most dangerous bird." Despite this, cassowaries are shy and elusive, avoiding conflict when possible. They feed primarily on fallen fruit, playing a crucial ecological role in seed dispersal and forest regeneration, especially for large-fruited trees. These ancient birds—relatives of the emu and ostrich—are like time travelers from a lost world, embodying a raw, grounded kind of magic. In both myth and biology, the cassowary teaches us to respect boundaries, honor the primal world, and walk softly with great strength.

Chalkydri
(Jewish Mythology)

celestial wisdom

The Chalkydri (sometimes spelled Chalkedri or Chalkydra) is a mythological, radiant bird described in Jewish apocryphal texts, particularly in the Second Book of Enoch, a work of mystical vision likely written between the 1st and 3rd centuries CE. In these ancient writings, the Chalkydri are said to stand before the throne of God, shining with fiery brilliance, with wings like flame and faces that radiate the light of the sun. They sing praises in the morning and at sunset, acting as divine heralds and eternal worshippers. The name is thought to be related to the Greek word chalkos, meaning bronze or copper, perhaps referring to their shining, metallic appearance. As angelic beings or "heavenly birds," they are symbols of divine illumination, eternal truth, and the celestial order. Their presence in the cosmic hierarchy situates them as mediators between heaven and Earth, and their unending songs echo the mystical notion that all of creation worships in harmony with the divine.

The Chalkydri are not tied to a physical habitat on Earth, but rather occupy a place in the heavenly firmament, somewhere between the angelic hosts and the sun itself. In the visionary cosmology of early Jewish mysticism—especially within Merkabah (chariot) mysticism—they represent the fiery, unknowable aspects of God's will, embodied in living form. These radiant birds appear alongside other celestial beings such as seraphim, ophanim, and cherubim, making them part of the divine entourage that carries out heavenly tasks and guards sacred mysteries. Because of their blazing radiance and sacred songs, the Chalkydri have been interpreted in later mystical and esoteric traditions as symbols of enlightenment, divine transformation, and the transcendence of the soul. While largely forgotten in mainstream theology, they remain a stunning example of how the imagination of early mystics visualized not just heaven, but the eternal harmony of the universe—sung, feathered, and aflame.

Chickadee
(Poecile atricapillus)

bringer of warnings and truth

The chickadee is a symbol of cheerfulness, curiosity, adaptability, and truth spoken with simplicity. In Native American traditions, particularly among the Cherokee and other Eastern Woodlands tribes, the chickadee is considered a bringer of warnings and truth, often appearing when it's time to pay attention to small details or trust one's instincts. Its cheerful song and curious nature have made it a symbol of lightness, bravery despite size, and finding joy in the everyday. Chickadees are known to warn other animals of danger through complex vocalizations, making them a symbol of honest communication and communal care. Because they are so active and inquisitive, they also represent playfulness, mental agility, and the ability to face darkness (like winter) with a light spirit.

Chickadees are native to North America, with the black-capped chickadee being one of the most familiar species across Canada and the northern United States. They thrive in cool temperate forests, mixed woodlands, and suburban areas, adapting easily to changing environments. These birds are highly social and intelligent, often storing food for later use and remembering hiding spots with impressive accuracy. They are cavity nesters, frequently using old woodpecker holes or birdhouses, and they remain in their home range year-round, even through harsh winters. Their distinctive "chicka-dee-dee-dee" call is not only charming—it's a complex form of communication that varies depending on the type of predator nearby. Despite their tiny size, chickadees are resilient and resourceful, embodying the idea that joy, truth, and strength often come in the smallest, most unassuming packages.

Chickcharney
(Bahamian Mythology)

luck

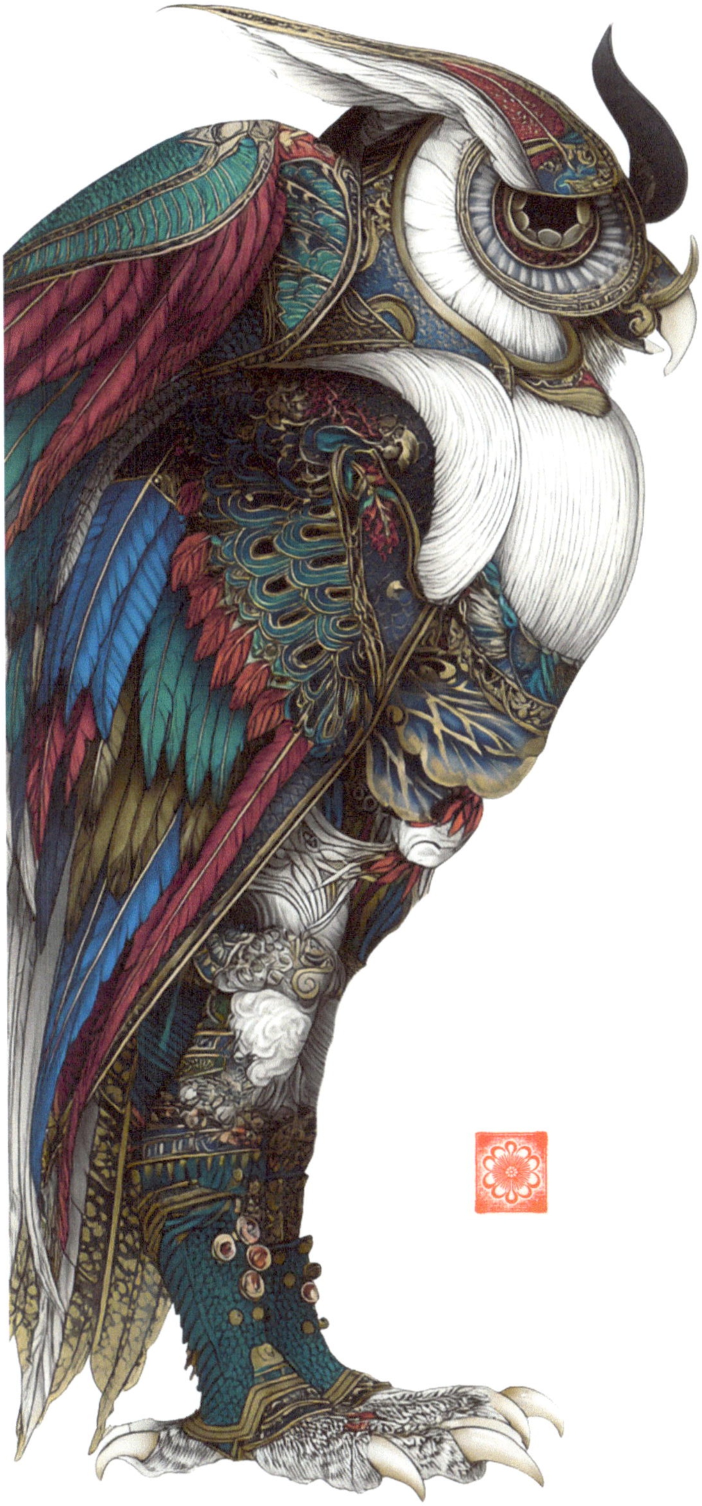

The Chickcharney is a creature of Bahamian folklore, said to inhabit the dense pine forests of Andros Island, the largest and most rugged island in the Bahamas. Described as a three-foot-tall, owl-like being, with shaggy feathers, piercing eyes, and the uncanny ability to twist its head completely around, the Chickcharney blurs the line between bird and spirit. Some accounts say it has three toes, others give it monkey-like traits, adding to its eerie charm. According to legend, the Chickcharney is a guardian of the woods—a bringer of either great fortune or bad luck, depending entirely on how it's treated. Show kindness and respect when encountering one, and it may reward you with success and good omens. Mistreat it, laugh at it, or show cruelty, and you risk misfortune, curses, or a lifetime of ill luck. This duality places the Chickcharney in the archetype of a trickster or guardian spirit, like many forest-dwelling beings in global folklore, teaching lessons about respect for nature and the unseen energies within it.

Though considered mythological, some believe the story of the Chickcharney may have been inspired by the now-extinct Tyto pollens, a species of large owl that once lived in the Bahamas. This ancient owl, larger than any living barn owl, may have survived in oral memory as a forest creature both real and supernatural. The Chickcharney's legend reflects the animistic worldview often found in Caribbean and African diaspora cultures, where the environment is alive with spirits, and actions toward nature have real, spiritual consequences. These stories continue to be shared in Bahamian communities today—not only as cautionary tales, but as cultural touchstones that emphasize the sacred bond between people, land, and the mysteries that dwell in the shadows of the forest. Whether cryptid, ghost, or metaphor, the Chickcharney endures as a symbol of the thin veil between our world and the magical one just beyond the trees.

Chough
(Pyrrhocorax pyrrhocorax)

divine intervention

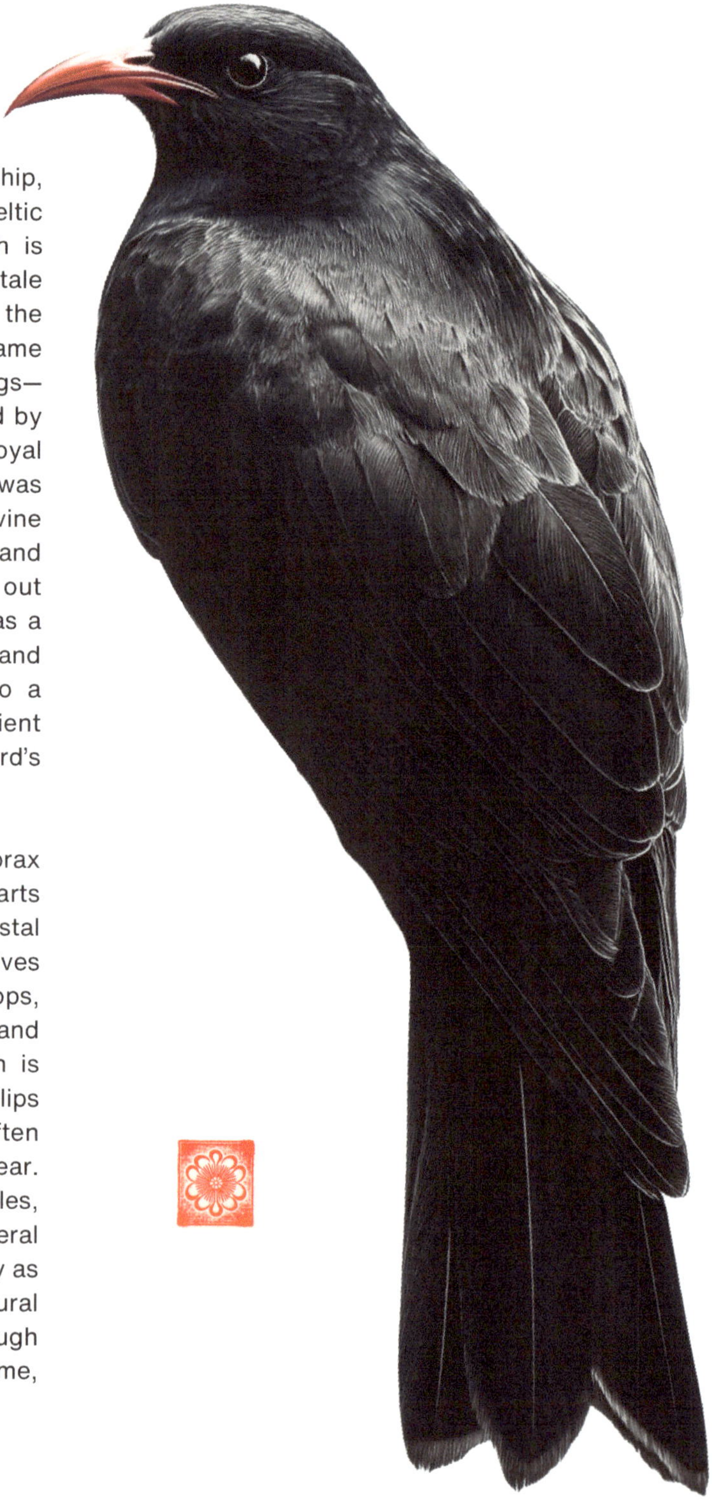

The chough is associated with sacredness, kingship, resurrection, and protection of ancient wisdom. In Celtic mythology and later Arthurian legend, the chough is deeply connected to King Arthur himself. One tale claims that after his death, Arthur's soul entered the body of a chough, and so in parts of Cornwall, it became taboo to harm the bird. Its striking red beak and legs—unusual among corvids—were believed to be stained by Arthur's blood, making the bird a living relic of royal sacrifice and heroism. In Christian symbolism, it was sometimes associated with monastic life and divine watchfulness, believed to nest near sacred sites and abbeys. Because it often lives high on cliffs and calls out with sharp, clear cries, the chough has been seen as a guardian of thresholds—between land and sea, life and death, the mortal world and the spiritual. It's also a symbol of rebirth, especially in folklore tied to ancient lands like Cornwall, Wales, and Brittany, where the bird's return each spring was celebrated as a hopeful sign.

In the real world, the red-billed chough (Pyrrhocorax pyrrhocorax) is native to Europe, North Africa, and parts of Central and South Asia, especially favoring coastal cliffs, mountainous regions, and open uplands. It thrives in cool to temperate climates with rocky outcrops, where it can nest in crags and feed on insects and larvae hidden in short grass and soil. The chough is known for its aerial agility, often seen performing flips and glides along cliff edges. It mates for life, often returning to the same nesting sites year after year. Though once in decline in parts of the British Isles, conservation efforts have led to its symbolic and literal return to areas like Cornwall, where it is seen not only as a rare bird but as a resurrected emblem of cultural identity. Whether through folklore or flight, the chough remains a keeper of legends and a witness to time, bridging myth and landscape with every wingbeat.

Cinnamon Bird
(Greek Mythology)

exoticism

The Cinnamon Bird, or cinnamologus, is a mythical creature believed by ancient Greek and Roman naturalists to reside in the remote and exotic lands of Arabia or South Asia, where cinnamon was harvested. According to the lore, these birds gathered sticks of cinnamon to build their nests high on inaccessible cliffs or treetops, beyond the reach of any man. In order to harvest the precious spice, humans devised clever strategies—one version involves shooting down the nests with lead-weighted arrows; another describes merchants tricking the birds by luring them with heavy meat, which the birds carried back to their nests, causing them to collapse under the weight and fall to the ground. The Cinnamon Bird thus became a symbol of mystery, abundance, and the clever balance between nature's gifts and human ingenuity. It also carried connotations of wealth, trade, and the allure of the unknown, representing the extravagant and often mystical perception of Eastern lands in classical and medieval Europe.

Of course, the Cinnamon Bird has no counterpart in the natural world—there is no species that builds nests of cinnamon—but the myth reflects a time when rare and luxurious spices were so prized they demanded mythological origin stories. Cinnamon, highly valued for its flavor, fragrance, and medicinal properties, was traded across vast desert and sea routes, shrouded in secrecy by those who wished to control its value. This fantastical bird helped explain why cinnamon was so difficult to obtain and so costly, while also elevating it to a near-sacred substance, guarded by mythical beings in distant, enchanted lands. In symbolic terms, the Cinnamon Bird can be seen as a metaphor for the hidden treasures of the Earth, the seductive mystery of faraway places, and the clever human desire to unlock nature's wonders—even if it means dreaming them into being.

Cockatoo
(Cacatua galerita)

messenger of the spirit world

The cockatoo is a powerful symbol of illumination, clarity, and expressive energy. In Aboriginal Australian mythology, the cockatoo is often seen as a messenger of the spirit world, a bird that brings guidance through its calls and radiant plumage. White cockatoos, in particular, are associated with light and purification, while pink or sulfur-crested cockatoos symbolize creative fire and emotional freedom. Their ability to mimic human speech and sounds has made them metaphors for conscious communication, the power of voice, and the need to speak one's truth. Because of their social nature and complex behaviors, cockatoos are also totems of community, loyalty, and emotional intelligence—birds that reflect the energy around them and remind us to be mindful of what we project. In esoteric traditions, the cockatoo is thought to activate the crown chakra, encouraging enlightenment and the embrace of higher awareness, especially in times of transition or creative blocks.

Native to Australia, New Guinea, and parts of Indonesia and the Philippines, cockatoos thrive in tropical and subtropical forests, woodlands, and even arid scrublands. They are highly adaptive and often live in flocks, relying on communal bonds for survival. Most recognizable are the sulfur-crested cockatoos (Cacatua galerita), with their expressive yellow crests, loud vocalizations, and playful nature. Their powerful beaks allow them to crack seeds, dig into tree bark, and even dismantle man-made structures—an echo of their symbolic connection to breaking down obstacles and challenging what is rigid or artificial. These birds are also among the most intelligent of parrots, capable of learning routines, mimicking voices, and forming deep bonds with their human companions. Their vibrant energy and emotional expressiveness make them a living emblem of freedom, individuality, and the radiant chaos of being fully alive.

Cockatrice
(English Folklore)

death and destruction

The cockatrice is a mythical beast said to hatch from a rooster's egg, incubated by a toad or serpent—a deeply unnatural beginning that immediately marks it as an abomination. This bizarre origin gave rise to its symbolic association with perversion of nature, cursed knowledge, and fatal arrogance. It was believed that the cockatrice could kill with a single glance or breath, turning people or animals to stone. Some legends added that even its reflection was deadly, placing it in the same deadly gaze mythology as the basilisk, with whom it's often conflated. In medieval Europe, particularly in England and France, the cockatrice became a common figure in bestiaries, where it was used to illustrate the spiritual danger of pride and sin—especially those that twisted the natural order. Alchemists also referenced it as a cautionary symbol in their metaphysical writings, often associating it with the corruptive side of transformation, the kind of creation born from impurity or ego rather than divine inspiration
.

Of course, the cockatrice never existed in the natural world—it is a product of fear, folklore, and the mysterious gaps in medieval understanding of biology. But its legend may have been inspired by misidentified fossils, sightings of strange lizards or snakes, or the psychological need to explain sudden illness and death. It was said that the only creatures immune to its gaze were weasels or roosters, and the crow of a real rooster could kill it—an ironic twist, given its supposed origin. The cockatrice also found a place in heraldry, often used by noble houses that wanted to display fearsome strength or divine protection against evil. Today, while the myth has faded into fantasy fiction and role-playing games, its core symbolism endures: the cockatrice represents what happens when power is twisted, when nature is defied, and when creation becomes destruction. It is a warning, a marvel, and a shadow of humanity's fascination with its own limits.

Common Loon
(Gavia immer)

messenger between worlds

The common loon holds deep symbolic meaning across Indigenous North American cultures, particularly among the Ojibwe, Cree, and Inuit peoples. It is often seen as a messenger between worlds, a creature that travels effortlessly between air, water, and the mystery below. The loon's mournful, echoing call is believed by many to be a voice of the ancestors, a guide for dreams, or a signal of deep inner change. In creation stories, such as the Anishinaabe flood myth, the loon plays a vital role in shaping the world, diving to retrieve earth from beneath the water to help recreate the land. Its ability to disappear beneath the surface and reappear far away makes it a symbol of introspection, emotional depth, and the unseen parts of the soul. In modern symbolism, especially in literature and music, the loon often represents solitude, longing, and the haunting beauty of wild places untouched by time.

The common loon is native to North America, breeding primarily in the lakes of Canada and the northern United States, particularly in regions like Minnesota, Maine, and the Adirondacks. It winters along coastal waters in the U.S., Mexico, and parts of Europe, preferring cool, clean freshwater lakes during nesting season and marine coastlines during the colder months. The loon is built for life in water—it has solid bones (unlike most birds) to help it dive, and it can swim underwater for long distances in search of fish and invertebrates. Its red eyes aid vision below the surface, while its legs, set far back on its body, make it a clumsy walker but an extraordinary diver. Loons are monogamous and highly territorial during breeding, often returning to the same nesting sites year after year. Their presence is a hallmark of pristine waters and healthy ecosystems. With their striking black-and-white plumage, low silhouettes, and otherworldly calls, loons embody the stillness and mystery of deep water, inviting us to reflect on what lies beneath the surface of our own thoughts and lives.

Cormorant
(Phalacrocorax carbo)

symbol of cooperation

The cormorant has long been a symbol of duality—both wealth and avarice, wisdom and darkness. In European folklore and Christian allegory, it was sometimes cast in a negative light, associated with greed because of its voracious appetite for fish. John Milton famously used it in Paradise Lost, where Satan appears in the form of a cormorant perched atop the Tree of Life, reinforcing its image as a cunning, shadowed figure. Yet in stark contrast, in Chinese and Japanese traditions, the cormorant is seen as a noble, skilled companion, trained for centuries by fishermen in a practice known as cormorant fishing. With a ring placed gently around its neck to prevent it from swallowing large fish, the bird dives and retrieves the catch, working in harmony with its human partner. In these cultures, it is a symbol of cooperation, mastery of the natural world, and an almost mystical bond between man and animal. So while in the West the cormorant might be viewed with suspicion, in the East it stands as a creature of elegance, precision, and partnership.

Cormorants are found on every continent except Antarctica, thriving along coastlines, rivers, lakes, and wetlands. The great cormorant (Phalacrocorax carbo), the most widespread species, is a master diver, able to plunge deep underwater and stay submerged longer than most aquatic birds. It thrives in temperate to subtropical climates, often nesting in colonies on cliffs, trees, or even man-made structures. Its dark, oily feathers, which lack the waterproofing seen in many other water birds, allow it to dive easily—but must be dried afterward, giving rise to its iconic, almost theatrical posture: wings outstretched in the sun, like a cruciform figure in meditation. Cormorants are intelligent, highly adaptable, and often misunderstood. Their ability to dive deep for what they need—and then surface again to reflect—makes them potent symbols of insight, hidden wisdom, and the balance between shadow and purpose.

Corncrake
(Crex crex)

secrecy

The corncrake is a bird steeped in folkloric nostalgia, especially in Celtic and Gaelic traditions, where its haunting, rasping call—once common across meadows and farmland—heralded the arrival of spring and the turning of the seasons. Yet, because it is rarely seen and so easily overlooked, it became a symbol of mystery, invisibility, and the unseen truths that linger just beneath the surface of ordinary life. In Irish and Scottish lore, the corncrake has been associated with the spirit world and ancestral voices, its call said to echo from the past. In many rural traditions, it was also a harbinger of change or farewells, and as the bird became increasingly rare in modern times, it came to embody vanishing beauty and the quiet erosion of the old ways. In literature and poetry, the corncrake often appears as a symbol of longing, retreat, and the poignant connection between memory and place.

Biologically, the corncrake is a member of the rail family, native to Europe and western Asia, with a breeding range that once extended across meadows, hay fields, and grasslands, especially in the British Isles, Ireland, and Scandinavia. It prefers temperate climates during breeding and migrates to sub-Saharan Africa for the winter, a long journey that reflects its deep connection to seasonal rhythms. The corncrake is secretive, living almost entirely hidden within tall vegetation, where it moves with a quiet, shuffling gait. Its raspy, mechanical "crex crex" call—repeated at dusk and into the night—is its most recognizable trait. Despite being a poor flier, it travels incredible distances during migration, a testament to its quiet strength. Due to habitat loss and changes in farming practices, corncrake numbers declined drastically in the 20th century, making it a focus of conservation efforts. Today, it stands as a symbol of fragility, the importance of stewardship, and the beauty of the voices we rarely notice until they are nearly gone.

Crane
(Grus grus)

good fortune

The crane is one of the most powerful and universally revered birds in world mythology, especially in East Asian, Indigenous American, and European traditions. In Japanese culture, the crane (tsuru) is considered a sacred creature, symbolizing good fortune, long life, and marital fidelity. It is said that cranes live for a thousand years, and folding one thousand origami cranes (senbazuru) has become a gesture of deep healing, hope, and prayer—especially for peace and health. In Chinese mythology, cranes are celestial beings that carry souls to heaven and accompany immortals, representing nobility, wisdom, and a harmonious life. Meanwhile, in Greek legend, cranes were associated with vigilance and prophetic insight, said to dance in celebration of the seasons and divine cycles. In many Native American traditions, particularly among the Plains and Pueblo peoples, cranes are seen as messengers and guardians of sacred knowledge, embodying grace, balance, and the sacred rhythm of nature. Across all these interpretations, the crane emerges as a symbol of elevation—physical, spiritual, and emotional.

There are around 15 species of cranes found across the globe, with notable ones including the common crane (Grus grus), whooping crane (Grus americana), and the red-crowned crane (Grus japonensis). Cranes are native to Eurasia, North America, and parts of Africa and Australia, with habitats ranging from wetlands and marshes to grasslands and tundra. Most species prefer temperate to cold climates and undertake spectacular migrations, forming dramatic V-shaped formations in the sky. Cranes are known for their elaborate courtship dances, long-term pair bonds, and haunting, trumpet-like calls that can carry for miles. They nest in shallow waters and are incredibly attentive parents. Because of habitat loss and environmental threats, many crane species are now vulnerable or endangered, and international efforts are underway to preserve both their environments and their cultural legacy. As living embodiments of elegance, devotion, and higher wisdom, cranes remind us to move through the world with intention, to honor our connections, and to rise above the chaos with quiet dignity.

Crow
(Corvus brachyrhynchos)

omen of change

Across world mythologies, the crow is a figure of duality and depth, often representing death and rebirth, intelligence, prophecy, and transformation. In Native American traditions, particularly among the Haida, Hopi, and Lakota peoples, the crow appears as a trickster spirit—clever, mischievous, and even creator-like in some stories. Among the Celts, crows were associated with war and the battlefield, often seen as the familiars of warrior goddesses like Morrígan, who could shape-shift into a crow and fly over battles. In Norse mythology, Odin, the god of wisdom, was accompanied by two ravens—Huginn and Muninn—but crows also shared this symbolism of thought, memory, and insight. In modern spiritual thought, crows are seen as harbingers of transformation, calling attention to cycles of change, the need to confront truth, and the presence of hidden knowledge. Because they are scavengers, they also represent purification through decay, the ability to reclaim what is forgotten, and the sacred role of endings in the greater arc of renewal.

Crows are found all over the world, with the American crow (Corvus brachyrhynchos), carrion crow (Corvus corone), and house crow (Corvus splendens) among the most widespread species. They inhabit a wide range of environments, from rural fields and forests to urban neighborhoods, and are highly adaptable to human presence. Crows thrive in temperate to subtropical climates, and their social intelligence is extraordinary: they recognize individual human faces, use tools, solve complex problems, and even engage in what appears to be ritualistic behavior—such as "funerals" for their dead. They mate for life, live in tight-knit family groups, and communicate with an intricate system of calls. Their glossy black feathers, often interpreted as ominous, actually shine with hues of blue and violet in the right light—mirroring the symbolism that beneath shadow, there is complexity, beauty, and meaning. Whether seen as omens, guides, or clever mischief-makers, crows are ultimately a reminder to see beyond appearances, embrace intelligence with humility, and honor the magic hidden in plain sight.

Cuckoo
(Cuculus canorus)

prophetic meaning

The cuckoo is a bird steeped in layered symbolism, much of it derived from its unusual behavior and eerie, unmistakable call. In many European folk traditions, the first call of the cuckoo in spring was believed to mark the turning of the seasons and held prophetic meaning—especially related to love, longevity, and future fortune. People would ask, “How many years do I have left to live?” and then count the number of cuckoo calls they heard. Yet, despite this hopeful role, the cuckoo has also long been a symbol of infidelity and deception, primarily because of its brood parasitism—the practice of laying its eggs in the nests of other birds, tricking them into raising its young. In literature and legend, this behavior has earned it associations with unfaithfulness, abandonment, and selfishness, as well as the darker, more cunning sides of survival. In Shakespearean works and medieval ballads, the cuckoo is often invoked as a sign of betrayal in love or deceit in friendships, yet it also represents the inevitability of cycles—things that return whether welcome or not.

The common cuckoo (Cuculus canorus) is native to Europe, Asia, and parts of North Africa, and migrates annually to sub-Saharan Africa for the winter. It thrives in a range of temperate habitats, including woodlands, open countryside, and farmlands, especially where host species such as warblers and pipits are plentiful. What makes the cuckoo so remarkable is its ability to mimic the eggs of its host species, as well as its biological programming to push out the other chicks once it hatches—ensuring it receives all of the care and food from its unsuspecting adoptive parents. Despite its reputation, the cuckoo plays an important role in the ecosystem, helping regulate insect populations and enriching avian biodiversity through its complex interactions. Its call, a distinct and mournful two-note song, has become one of the most evocative sounds in nature—a reminder of impermanence, the passing of time, and the often uncomfortable truths that lie beneath seemingly simple melodies.

Cuckoo Roller
(Leptosomus discolor)

guardians of balance

Symbolically, the Cuckoo Roller represents change, inner balance, and the ability to see from new angles. Found in the dense forests of Madagascar and the Comoros, this solitary bird is often associated with hidden wisdom and introspection, due in part to its elusive behavior and curious flight pattern. Its slow, looping glides and rolling aerial maneuvers have earned it its name, suggesting a certain fluidity, unpredictability, and playful defiance of gravity. While it doesn't carry heavy mythological baggage, in Malagasy culture, many forest birds—including the cuckoo roller—are regarded with quiet reverence as part of the forest's living soul, and often treated as messengers or guardians of balance. Because of its distinctive and solitary nature, the cuckoo roller can also be interpreted symbolically as a creature of personal transformation, emotional detachment, and graceful navigation through transitional spaces—a reminder to embrace change with grace, even when the path is uncertain.

Biologically, the cuckoo roller is an evolutionary outlier, being the sole member of both its genus (Leptosomus) and its own family (Leptosomidae). Found almost exclusively in humid, evergreen forests, it thrives in tropical climates, often perching motionless for long periods before making short, fluttery flights to catch insects, chameleons, and small reptiles. It is sexually dimorphic—males have a pale gray-green head and chest with a bronzy back, while females are more brown and streaked—making them one of the few bird species with such visible gender differences. Despite being somewhat sluggish in demeanor, their aerial abilities are striking, and their haunting, echoing call can be heard reverberating through the forest canopy. Their adaptability to a variety of forest types, from primary rainforest to secondary woodland, speaks to their symbolic role as quiet navigators of change, moving between the visible and hidden with ease and reminding us to observe before acting, and to honor the quiet power of transition.

Curlew
(Numenius arquata)

guide of souls

The curlew has long been a creature of poetic sorrow and spiritual depth. In Celtic mythology, it is associated with the Otherworld, often believed to be a bird that exists between realms, a guide of souls or a symbol of grief, longing, and transformation. Its call—an eerie, spiraling whistle—has been described in folklore and literature as a sound of mourning, echoing across empty fields like a voice from the past. In some Scottish and Irish tales, the curlew's cry was considered an omen of death or a lament for those lost, especially for soldiers or those who died far from home. Yet, it is not a symbol of fear—it represents the gentle sadness of change, the solitude of wide-open spaces, and the importance of listening to silence. In modern poetry and ecological writing, the curlew has become a symbol of endangered beauty, as many species face steep population declines due to habitat loss.

The Eurasian curlew (Numenius arquata) is the largest wading bird in Europe, instantly recognizable by its long, downward-curving bill, used to probe deep into soft mud for worms and invertebrates. It is native to Europe and Asia, favoring open wetlands, estuaries, moorlands, and coastal meadows, with migratory populations wintering as far south as North Africa and the Middle East. It thrives in cool, damp climates, and depends on intact peatlands and mudflats—habitats that are increasingly threatened. The curlew is a ground-nester, vulnerable to disturbance and predation, and its eerie song is most often heard during the breeding season, rising into the dawn or dusk air like a spectral echo. Its presence is cherished in conservation circles as a symbol of what is vanishing and worth saving, and efforts to protect curlew habitats have become emblematic of broader movements to defend biodiversity. The curlew, with its haunting voice and timeless grace, reminds us that not all beauty is loud—and some of the most meaningful songs are the ones that sound like silence turning into memory.

Dove
(Columba livia)

peace and love

The dove is perhaps the most widely recognized bird in symbolic language, embodying themes of peace, innocence, hope, and spiritual grace. In Christianity, the dove represents the Holy Spirit, famously descending from heaven during the baptism of Christ, and appearing in religious art as a sign of divine communication, purity, and sacred love. In Judaism, a dove returned to Noah with an olive branch, signaling the end of the flood and the renewal of the Earth—this story is the root of the now-universal symbol of the dove with an olive branch as a sign of peace. In Greek mythology, doves were sacred to Aphrodite, the goddess of love, reinforcing their link to beauty, romance, and feminine energy. In Islam, it is believed a dove whispered into the ear of the Prophet Muhammad, carrying the wisdom of Allah. Doves are also symbols of the soul in flight, representing the transition from the earthly to the spiritual realm, especially in funerary art and mystic poetry. Their soft cooing and gentle demeanor reinforce their symbolism as messengers of serenity, healing, and eternal devotion.

Biologically, doves are found almost everywhere in the world, with the rock dove (Columba livia)—the ancestor of the domestic pigeon—being one of the most widespread. Doves thrive in a wide range of climates, from urban cities and Mediterranean coasts to forests, deserts, and farmland. They are highly adaptable and are known for their monogamous bonds—pairs often mate for life, building simple nests and co-parenting their chicks with great attentiveness. Their plumage ranges from radiant whites and silvery grays to soft browns and rose hues, depending on species. Doves feed mostly on seeds and fruits and are gentle in both behavior and flight, often gliding rather than darting. Their presence at weddings, ceremonies, and public peace events has elevated them into real-world emissaries of symbolic hope. Whether released in joy or mourned in poetry, the dove reminds us to seek peace, to live with compassion, and to listen for the quiet grace that underlies all things.

Drongo
(*Dicrurus macrocercus*)

guardian of the land

The drongo is best known for its fierce and fearless behavior, often challenging larger birds of prey to defend its territory and even chasing them off despite being a fraction of their size. In Aboriginal Australian mythology, particularly among the Northern Territory's Yolngu people, the willie wagtail (often associated or conflated with drongos) is a trickster figure, while in broader Indigenous belief, the drongo is respected as a protector bird—a guardian of the bush who watches over the land with sharp eyes and sharper calls. In African folklore, the drongo is often portrayed as a clever deceiver who uses mimicry to trick other animals out of their food—a behavior that's been scientifically observed. Its mimicry isn't just symbolic—it's literal: the drongo can imitate the alarm calls of other species to scare them away and snatch their meal, earning it the nickname "the feathered con artist." As a result, it has come to represent the power of wit over brawn, adaptability in the face of challenge, and survival through cleverness rather than force.

The drongo family includes more than 20 species, found throughout Africa, southern Asia, and Australia, with habitats ranging from forests and savannas to agricultural land and open woodland. The most well-known species include the fork-tailed drongo (Dicrurus adsimilis) of Africa and the spangled drongo (Dicrurus bracteatus) of Australia. They thrive in warm and tropical climates, are insectivorous, and are often seen swooping gracefully through the air in pursuit of prey. Their striking black plumage sometimes shimmers with a blue or green sheen in sunlight, adding to their visual mystique. Though solitary or found in pairs, drongos are incredibly social within their ecosystems, often interacting with other species through mimicry, alarms, and cooperative warning systems. Symbolically, they embody strategic thinking, fierce independence, and the idea that the smallest voice can still command a great deal of power.

Duck
(Anas platyrhynchos)

emotional balance

In symbolism, the duck is often associated with emotional balance, adaptability, and family devotion. Ducks can move seamlessly between land, water, and sky, making them natural symbols of fluidity between worlds—the physical, emotional, and spiritual. In Native American traditions, ducks are seen as bringers of water and fertility, often connected to the nurturing qualities of mothers and the life-giving properties of wetlands. In Chinese folklore, a pair of mandarin ducks symbolizes romantic fidelity and enduring love, often given as a wedding gift to represent a harmonious marriage. In Celtic and Norse symbolism, ducks were seen as guardians of the otherworldly lakes and marshlands, moving through liminal spaces and connecting people to hidden truths. Their ability to remain calm on the surface while paddling furiously below is frequently used as a metaphor for grace under pressure and emotional depth hidden beneath a composed exterior.

Biologically, ducks are found almost everywhere on Earth, with mallards (Anas platyrhynchos) being one of the most widespread and well-known species. They thrive in wetlands, lakes, rivers, and ponds, and are highly adaptable to a range of climates, from arctic tundras to tropical lagoons. Ducks are omnivores and opportunists, feeding on aquatic plants, insects, and small invertebrates. They are strong fliers, excellent swimmers, and nurturing parents—many species are fiercely protective of their young. Male ducks often display vibrant, iridescent colors during the breeding season, while females are typically more camouflaged to protect their nests. Ducks' down feathers also symbolize comfort, warmth, and softness, and they are one of the few birds whose relationship with humans spans from wild to deeply domestic. As symbols, ducks remind us to stay emotionally buoyant, protect what matters, and find peace in the everyday rhythms of life.

Eagle
(Aquila chrysaetos)

divine connection

The eagle is one of the most revered symbols in global mythology, standing for power, freedom, vision, and divine connection. In Native American cultures, the eagle is considered a sacred messenger between humans and the Creator, with its feathers used in ceremonies and spiritual rituals as conduits of prayer and blessing. In ancient Rome, the eagle was the symbol of Jupiter and the Roman legions—a sign of imperial power, authority, and the favor of the gods. In Christianity, it represents spiritual ascension and the resurrection, and in many shamanic traditions, it is seen as the embodiment of the sun, of higher truth, and the ability to see clearly from great heights. The eagle's flight is a metaphor for rising above earthly concerns, accessing wisdom, and acting with precision and courage. It is often called upon as a symbol of leadership, nobility, and destiny.

Eagles are found across every continent except Antarctica, with iconic species like the bald eagle (Haliaeetus leucocephalus) in North America and the golden eagle (Aquila chrysaetos) in Europe, Asia, and parts of Africa. They thrive in mountains, coastal regions, river valleys, and open plains, preferring temperate to arid climates. Eagles are apex predators, known for their incredible vision, powerful talons, and strategic hunting behaviors. They build massive nests (eyries) in high places, often returning to the same site year after year. Mated pairs are monogamous and share in raising their young. Watching an eagle soar evokes awe—it embodies grace in power, stillness in motion, and the silent confidence of a being that knows its purpose. In every tradition, the eagle reminds us to rise above distraction, to trust our inner sight, and to lead with courage and clarity.

Eagle Owl
(Bubo bubo)

insight

The eagle owl is steeped in symbolism tied to secrecy, intuition, and the nocturnal forces of insight and protection. In European and Eurasian folklore, this owl was often regarded as a guardian of ancient knowledge, appearing in tales as a silent observer of human folly and a keeper of secrets best left undisturbed. Unlike the smaller barn owl, which is often linked to sorrow or ghostly omens, the eagle owl commands a more sovereign, almost shamanic energy—symbolizing the ability to see through deception, navigate darkness, and act with silent strength. In some Slavic traditions, it was believed that hearing an eagle owl near your home was a message from the spirit world, sometimes a warning, other times a reassurance that an unseen force was watching over you. Its intense stare and horn-like ear tufts give it an otherworldly appearance, often making it a totem of hidden truths, ancestral knowledge, and psychic power.

The Eurasian eagle owl is native to Europe, North Africa, and vast stretches of Asia, from Spain and Scandinavia to the steppes of Mongolia and the mountains of Iran. It thrives in varied climates and terrains, including forests, rocky outcrops, deserts, and high-altitude cliffs. It is a top nocturnal predator, capable of taking down prey as large as foxes and small deer, though it typically feeds on rabbits, rodents, and birds. Eagle owls have a wingspan of up to 6.5 feet (2 meters) and fly in complete silence, thanks to their specialized feathers. They are monogamous, nesting in cliff crevices or abandoned eagle nests, and fiercely territorial. What makes them especially compelling is the way they embody the paradox of power cloaked in silence—an apex hunter that moves unseen, a creature that listens more than it speaks. Symbolically, the eagle owl is a reminder to trust your perception, embrace solitude when needed, and act with deliberate, unwavering purpose.

Egret
(Ardea alba)

inner peace

The egret represents grace under pressure, inner peace, and sacred purpose. In many Asian traditions, particularly in Japanese and Chinese folklore, the egret is considered a sign of nobility, patience, and balance, often depicted in poetry and painting as a solitary figure amid reeds or mist, embodying a refined spirit unshaken by the chaos around it. In Native American symbolism, especially among southeastern tribes, egrets are seen as symbols of vigilance, self-reliance, and spiritual alignment, as they stand perfectly still for long periods before striking decisively—a lesson in acting only when the time is right. Their snowy white plumage connects them to purity, honesty, and the divine, often interpreted as messengers or guides in dream symbolism. Because they are wading birds, existing between land, water, and sky, they also represent the interconnectedness of the material and spiritual worlds, gliding effortlessly between what is seen and what is felt.

Egrets are found across every continent except Antarctica, with the great egret (Ardea alba) being one of the most widespread species. They thrive in wetlands, marshes, swamps, and shallow lakes, preferring temperate to tropical climates. Egrets are skilled hunters, feeding mostly on fish, frogs, and insects, which they catch with sudden, lightning-fast strikes after long periods of stillness. Their tall, slender bodies, S-curved necks, and delicate plumes—especially in breeding season—make them one of the most visually graceful birds. During the 19th century, egret feathers became highly sought after in the fashion industry, leading to mass slaughter and ultimately to the birth of the modern conservation movement. Today, the egret is often seen as a symbol of ecological awakening and protection, and its return to many once-depleted wetlands stands as a powerful testament to the resilience of both nature and spirit. It teaches us that true strength often lies in stillness, and clarity comes when we learn to move through life with poise and presence.

Falcon
(Falco peregrinus)

spiritual ascension

The falcon symbolizes vision, power, precision, and spiritual ascension. In Ancient Egyptian mythology, the falcon was sacred to Horus, the sky god whose eyes represented the sun and moon. As Horus took the form of a falcon, he became a symbol of kingship, divine authority, and celestial protection, with pharaohs often portrayed with falcon headdresses or companions. In medieval Europe, falconry became the sport of the nobility, and the falcon came to represent nobility of character, loyalty, and controlled aggression—an apex predator that served with grace and discipline. In Native American symbolism, particularly among plains tribes, the falcon is a messenger from the spirit world, representing clear insight, swift action, and fearlessness. Because falcons are known to hunt from great heights and strike with unerring accuracy, they are often linked to strategic thinking, timing, and the mastery of one's path—never wasting energy, always locked onto the right moment.

The most iconic species, the peregrine falcon (Falco peregrinus), is found on every continent except Antarctica and is famous for being the fastest animal on Earth, reaching diving speeds of over 240 miles per hour (386 km/h) during its hunting stoop. Falcons thrive in diverse habitats, from coastal cliffs and open tundras to urban skyscrapers, where they nest high and hunt from above. They prefer temperate to arid climates, and are skilled aerial hunters, feeding on smaller birds and mammals. Falcons are solitary, fiercely territorial during nesting, and incredibly loyal to their mates, returning to the same nest sites year after year. With their sharp eyesight, aerodynamic form, and stoic demeanor, falcons embody the idea of focused intention—the power of knowing your target and pursuing it with relentless clarity. Whether in myth or flight, the falcon is a call to rise above distraction, trust your instincts, and strike only when the moment is truly right.

Firebird
(Slavic folklore)

transformation

The Firebird is a mythical glowing bird found in the epic tales and fairy tales of Russian, Ukrainian, and wider Slavic folklore. Often described as having feathers that shimmer like gold and flame, with eyes like crystal and a tail that lights up the forest, the Firebird is not merely a prize—it is a catalyst for destiny. In these stories, the Firebird is often the object of a great quest. A prince may discover one of its feathers and become obsessed, or he may be commanded by a tsar to capture it, leading him on a path of hardship, wisdom, and magical transformation. The Firebird symbolizes the soul's calling, a shining force that compels a person to leave comfort behind and undergo a journey of spiritual growth. It is also a creature of paradox—a bringer of great reward and great peril—often causing turmoil wherever it is taken, suggesting that the divine or miraculous cannot be possessed without consequence. In this way, the Firebird is a symbol of divine light, uncontainable inspiration, and the danger of desire that outpaces wisdom.

While there is no real-world counterpart, the Firebird may have been inspired by sightings of exotic birds like peacocks, golden pheasants, or even flamingos, brought back from the East along trade routes and exaggerated in tales. Its myth spread far beyond Slavic lands, influencing ballet (notably Igor Stravinsky's The Firebird), opera, and fantasy literature, where it often appears as a phoenix-like figure. Symbolically, the Firebird has become a fusion of rebirth, radiant beauty, and the sacred spark of transformation, much like the phoenix of Greco-Roman myth. Its role in folklore serves as both an inner light to follow and a warning: that not all brilliance can—or should—be caged. The Firebird invites us to chase the extraordinary, but to do so with humility, courage, and respect for the magic we seek to touch.

Flamingo
(Phoenicopterus roseus)

beauty and grace

The flamingo carries rich symbolic associations across cultures, often representing beauty, transformation, and the balance between emotion and poise. In ancient Egyptian mythology, flamingos were associated with the sun god Ra and the color red—a sacred color linked to divine energy and rebirth. In Caribbean and South American folklore, the flamingo often appears as a symbol of celebration, sensuality, and vibrant self-expression, while in modern spiritual symbolism, it represents emotional clarity, standing tall in vulnerability, and thriving in environments that others may find unwelcoming. Its ability to stand calmly on one leg for hours has made it a metaphor for inner balance and mindfulness, while its bright pink plumage—derived from the carotenoids in its diet—offers a lesson in transformation through nourishment: we are what we absorb, emotionally and spiritually. Because flamingos are also highly social birds that live in massive colonies, they represent community, connection, and the beauty of synchronized living.

Flamingos are native to Africa, the Caribbean, South America, southern Europe, and parts of Asia, thriving in tropical and subtropical regions near saltwater lagoons, alkaline lakes, and estuarine wetlands. They favor extreme environments where few other species thrive, relying on their uniquely adapted beaks to filter-feed on algae, plankton, and tiny crustaceans. Their striking pink color—ranging from pale blush to electric coral—comes entirely from their diet, making their appearance a symbol of internal richness expressed outwardly. Flamingos are monogamous within a season and nest communally, building mud mounds to raise a single chick. Their elaborate courtship dances, involving synchronized head-flagging, wing salutes, and communal strutting, make them icons of graceful unity and expressive harmony. Whether seen as a symbol of joy, elegance, or soulful alignment, the flamingo invites us to honor our vibrant selves, nurture emotional equilibrium, and find strength in soft, steady beauty.

Frigatebird
(*Fregata magnificens*)

duality of strength

The frigatebird represents aerial dominance, strategic brilliance, and the duality of strength and subtlety. In Polynesian and Micronesian seafaring cultures, the frigatebird is revered as a guide and messenger, often appearing in legends as a navigator of the soul and harbinger of landfall, its silhouette etched against the sky after long voyages. It is sometimes seen as a link between the heavens and the ocean, sailing the air with such precision that it became a totem of divine movement and wind-borne wisdom. In more modern, Western symbolic thought, the frigatebird has come to embody pirate-like cleverness and agility, as it is infamous for stealing fish mid-air from other birds—a behavior called kleptoparasitism. This has earned it a reputation as a sky pirate, an opportunist that thrives not through brute strength but through strategy, speed, and timing. It reminds us that success can come not from force alone, but from knowing when to act and how to outmaneuver with elegance.

Native to tropical and subtropical oceans worldwide, particularly around the Galápagos Islands, the Caribbean, and the Pacific coasts of the Americas, frigatebirds are creatures of open air and wind currents. They rarely land on water, despite being seabirds, and are almost always seen soaring effortlessly on thermals, able to stay aloft for days. They thrive in hot, coastal climates, nesting in mangroves, rocky cliffs, and remote islets. With their long, forked tails and expansive wingspans (over 7 feet/2 meters), they are acrobats of the sky. The males are particularly striking during mating season, inflating a bright red throat pouch like a balloon to attract females—making them not only skilled hunters but also masters of flamboyant courtship. Frigatebirds feed on flying fish and squid, either caught in mid-air or stolen from other seabirds. Symbolically, they remind us to ride the currents, seize the moment, and balance boldness with precision, thriving not by overpowering others, but by using the wind—and wisdom—to our advantage.

Gagana
(Russian Folklore)

resilience and supernatural

The Gagana (sometimes Gagana the Bird or Iron-Beaked Gagana) is a creature of Slavic and Russian folklore, often portrayed as a guardian of the boundary between the known world and the supernatural. Though not as widely known as the Firebird or Zhar-Ptitsa, Gagana holds a more esoteric, powerful presence, symbolizing untamed magic, ancestral strength, and the key to unlocking hidden knowledge or forbidden spaces. She is typically described as a large, birdlike beast with a beak of iron and claws of copper, features that give her an almost elemental quality—part earth, part fire, part myth. Gagana is sometimes said to live on the mythical island of Buyan, a place that appears and vanishes like mist in Russian lore, guarded by fierce winds and invisible magic. In some versions, she serves as a gatekeeper or servant to Baba Yaga, the fearsome forest witch, and is said to guard sacred keys or magical thresholds, revealing herself only to those who are worthy or destined to cross them.

Though clearly mythological, Gagana's image evokes real-world raptors like eagles, vultures, or even griffin-like hybrids, with her iron beak and predatory talons lending her the air of something ancient, wild, and untouchable. She's not a bird of song or courtship, but of power, protection, and primal force—a creature to be respected, not tamed. Unlike the more radiant Firebird, Gagana isn't associated with beauty or light, but rather with the raw, often uncomfortable truth of transformation. In a symbolic sense, Gagana teaches that wisdom must be earned, often through trial, suffering, or confrontation with shadow. She may guard a magical object, a gateway, or a soul's destiny—but she will not part with it easily. Gagana is a reminder that the deepest truths are guarded by fire, metal, and the courage to face what others fear, and that real power often lies hidden in the wild places few dare to go.

Galah
(Eolophus roseicapilla)

emotional resilience

The galah holds a special place in Aboriginal Australian symbolism and modern Aussie culture. In traditional stories, it's sometimes seen as a spirit of laughter and light, a trickster figure that brings balance through disruption, challenging seriousness with playfulness. Its loud squawks, acrobatic flying, and social antics make it a symbol of self-expression, vitality, and emotional honesty. In modern Australian slang, calling someone a "galah" can mean they're being a bit of a clown—silly, bold, maybe even foolish—but it's often said with affection. The galah represents the freedom to be loud, to break norms, and to live joyfully, without apology. It's a bird that reminds us not to take life too seriously and to speak our truth, even if it ruffles feathers. Yet beneath the playful surface is a deeply emotional and intelligent creature, one that forms strong lifelong bonds and expresses genuine affection—making it also a symbol of loyalty, partnership, and emotional resilience.

Biologically, the galah is a species of cockatoo, native to almost every part of mainland Australia, thriving in open woodlands, grasslands, farmland, and even urban areas. With its bright pink chest and face, pale grey back and wings, and expressive crest, it's instantly recognizable—and impossible to ignore. Galahs are highly social, often seen in noisy flocks, chattering as they feed on seeds and grains or perform aerial tricks in the sky. They thrive in arid to temperate climates, and have adapted remarkably well to human-altered landscapes, even nesting in buildings and street trees. Galahs are monogamous and mate for life, displaying deep affection and cooperative parenting. In captivity, they are known for their intelligence and need for mental stimulation, reflecting their symbolism of curiosity, emotional depth, and playful rebellion. To encounter a galah is to be reminded that joy can be radical, love can be loud, and being a little wild might just be your superpower.

Gannet
(Morus bassanus)

spiritual nourishment

The gannet symbolizes commitment, focus, and spiritual nourishment. Its high-speed dives—sometimes from over 100 feet (30 meters) above the sea—make it a symbol of determination and calculated action, often interpreted as a totem for those who act decisively, fearlessly, and with purpose. In Celtic and Northern European seafaring traditions, the gannet was seen as a bringer of good fortune and a guardian of the sea, its appearance signaling rich fishing grounds or fair weather. Because it nests in massive, tightly packed colonies on remote cliffs and islands, it also embodies community, endurance, and harmony within chaos—a reminder that individuality can thrive within a collective. In some spiritual readings, the gannet represents a bridge between elements, moving seamlessly between the heavens and the ocean, symbolizing the soul's ability to plunge deep into emotional or subconscious waters and return with nourishment.

The northern gannet (Morus bassanus) is found in the North Atlantic, breeding in large colonies on cliffs in Scotland, Ireland, Iceland, Norway, and parts of eastern Canada, and wintering farther south in temperate and subtropical waters. It thrives in coastal marine environments, especially where cold currents drive nutrients to the surface and fish schools abound. Gannets are expert hunters, feeding primarily on fish such as mackerel and herring. They're known for their incredible diving technique: folding their wings back and plunging into the ocean at high speed, often penetrating deep below the surface. Gannets are monogamous and return to the same nest sites year after year, reinforcing their symbolism of fidelity, loyalty, and sacred routine. Their stark white plumage with golden-tinged heads, piercing blue eyes, and vast wingspans give them an almost angelic appearance in flight. Altogether, the gannet is a powerful reminder that grace and strength can coexist, and that plunging into the unknown can bring abundance when done with clarity and trust.

Garganey
(Spatula querquedula)

seasonal change

The garganey is a bird of movement, seasonal change, and sensitivity to nature's rhythms. As one of the earliest migrating ducks in Europe, it has been historically seen as a herald of spring, returning from its wintering grounds in Africa to breed in the wetlands of Europe and Asia. In various folklore traditions, migratory birds like the garganey are associated with messages from the spirit world, the flow of time, and the wisdom of knowing when to move on. Because it is a relatively shy and secretive bird, the garganey also carries connotations of grace in quietness, resilience without fanfare, and emotional attunement—a creature that navigates vast journeys without calling attention to itself. It reminds us to listen to the changing winds, to honor cycles, and to find strength in fluidity and instinct.

The garganey breeds across Europe and western Asia, favoring shallow freshwater wetlands, marshes, flooded meadows, and reed beds during the spring and summer. In autumn, it migrates to sub-Saharan Africa, India, and Southeast Asia, seeking warm climates with ample aquatic vegetation and insects. Males are particularly striking during the breeding season, with their bold white crescent above the eye, chestnut head, and patterned flanks, while females are more subtly marked for camouflage. Though small and lightweight, garganeys are strong, long-distance flyers, capable of covering thousands of miles between breeding and wintering grounds. They feed by dabbling—tipping forward in shallow water to forage on plants, seeds, and small aquatic creatures. Because they often inhabit peaceful, secluded places and prefer to stay hidden in vegetation, they embody the quiet, essential role of nurturing life at the edges of visibility. In essence, the garganey is a symbol of subtle transition, instinctual wisdom, and the sacred art of knowing when and where to journey.

Garrulus (Jay)
(Garrulus glandarius)

guardians of sacred knowledge

The jay represents intelligence, communication, secrecy, and the guardianship of sacred knowledge. In Celtic and European folklore, the jay was often seen as a trickster or forest messenger, its bold coloring and loud voice contrasting with its hidden and cautious behavior. Because it mimics the calls of other birds and even human-made sounds, the jay is associated with adaptability, impersonation, and the power of voice. It's considered a keeper of secrets—able to vanish quickly into the trees, rarely seen but often heard—making it a symbol of discretion, intelligence in action, and the power of observation. In myth and modern spiritual symbolism, jays are seen as beings that carry and protect acorns, linking them to memory, foresight, and the preservation of future potential. They teach that boldness and planning can coexist, and that true wisdom is often cloaked in color and song.

The Eurasian jay (Garrulus glandarius) is native to Europe, North Africa, and parts of Asia, and it thrives in temperate forests, especially oak woods, where it plays a vital ecological role as a seed disperser—particularly for acorns, which it hides in caches for winter. Many of these acorns are never retrieved, resulting in the natural propagation of oak trees, giving the jay a very literal connection to growth and forest regeneration. The jay's plumage is a striking mix of soft browns, black, white, and flashes of iridescent blue on its wings, making it both beautifully visible and well-camouflaged in the dappled light of woodland. It prefers temperate climates and often forages alone or in small family groups. Known for its complex social behavior and vocalizations, the jay's intelligence places it alongside crows and magpies in the corvid family. In both nature and myth, the jay is a powerful reminder to speak with intention, protect what matters, and trust the value of things not immediately seen.

Goshawk
(Accipiter gentilis)

loyalty

The goshawk is often associated with control, fierce loyalty, and a warrior's clarity of vision. In medieval Europe, especially in falconry traditions, the goshawk was seen as the falconer's bird of nobility and battle—bold, swift, and deeply attuned to its handler. It represented martial strength, precision, and the ability to strike without hesitation, and it was reserved for knights and nobles due to its temperament and skill. In Norse and Celtic traditions, goshawks were linked to spiritual guardianship and ancestral connection, often appearing as omens or totemic guides. Because it hunts in forests with incredible agility—able to weave between trees at high speed—it also symbolizes mastery of the environment, decisiveness under pressure, and the balance between aggression and discipline. In some Native American symbolism, hawks like the goshawk are messengers of insight, calling the observer to pay attention to subtle truths, to act when the time is right, and to strike with purpose rather than force.

The northern goshawk is native to temperate forests across North America, Europe, and Asia, preferring dense woodlands, boreal forests, and mountainous terrain where its powerful, short wings and long tail allow it to maneuver with deadly elegance. It thrives in cool to temperate climates and feeds on medium-sized birds and mammals such as grouse, pigeons, rabbits, and squirrels. The goshawk is a solitary and territorial hunter, known for ambush-style attacks, using stealth and sudden speed rather than soaring or pursuit. It is striking in appearance—broad-chested, with slate-gray feathers, piercing red or orange eyes, and a fierce, almost ancient presence. Fiercely protective of its nest and young, the goshawk is also a symbol of parental vigilance and generational strength. Whether seen as a creature of war, wisdom, or wild instinct, the goshawk reminds us to hone our inner edge, act with intention, and move through the world with fierce grace and unwavering clarity.

Grouse
(Tetraoninae – subfamily)

seasonal cycles

In folklore and symbolism, the grouse is often associated with earth energy, fertility, and the dance of survival. In Celtic and Indigenous North American traditions, grouse were revered for their courtship displays, particularly the rhythmic drumming or "lekking" dances performed by males to attract mates. These displays, conducted on the same sacred grounds year after year, have imbued the grouse with symbolism tied to ritual, seasonal cycles, and the power of intention and repetition. In Native American spiritual systems, especially among Plains tribes, the grouse dance was mimicked in ceremonial rites to connect with the natural world, invoking the bird's energy of balance, renewal, and attunement with the Earth. Symbolically, the grouse teaches the importance of tuning into life's natural pace, respecting personal space, and embracing rootedness rather than constant motion. It is also seen as a guardian of the threshold between the seen and unseen, a bird whose presence may be quiet—but whose spirit is steady and strong.

Grouse species—such as the ruffed grouse (Bonasa umbellus), sage grouse (Centrocercus urophasianus), and black grouse (Lyrurus tetrix)—are native to North America, Europe, and Asia, inhabiting woodlands, heathlands, tundra, and mountainous scrub. They thrive in cool, temperate to subarctic climates, where they blend seamlessly into their environment. Most are ground-dwelling birds, with mottled brown or grey plumage that offers excellent camouflage. They feed on buds, shoots, berries, and insects, and many species are sedentary, enduring even the harshest winters in snow-covered thickets. During spring, their courtship rituals involve strikingly rhythmic drumming, foot-stamping, and puffed-up displays—not for flight, but for rooted power and expression. Their commitment to place, cycle, and seasonal repetition reminds us that there is strength in tradition, in staying grounded, and in knowing when to move—but also when to simply dance where you are.

Guinea Fowl
(Numididae – family)

truth-tellers

In traditional West African symbolism, especially among the Yoruba and Ashanti peoples, the guinea fowl is revered as a sacred bird, often appearing in parables, proverbs, and spiritual stories. Its distinct, speckled plumage represents clarity and discernment—the ability to see through illusion and chaos. Guinea fowl are also seen as protectors and truth-tellers, believed to sense danger before others do and sound the alarm, making them symbols of watchfulness, protection, and the wisdom of collective awareness. In some myths, the guinea fowl is associated with the Orisha Obatala, a deity of purity and wisdom, who sends the bird as a messenger. Because they are fiercely loyal to their flock and exhibit strong social behavior, they are also totems of community, loyalty, and the power of shared vigilance. Their quiet strength reminds us that the flock is stronger than the individual, and that wisdom is often expressed not through grand gestures but through presence, persistence, and quiet authority.

Guinea fowl are native to sub-Saharan Africa, particularly savannas, woodlands, and semi-arid grasslands, though they have been widely domesticated and now live on farms and homesteads around the world. The most recognized species is the helmeted guinea fowl (Numida meleagris), known for its round body, featherless blue face, bony crest, and white-spotted black plumage. These birds thrive in warm, dry climates and are exceptional foragers, feeding on insects, seeds, and small reptiles. They move in tight flocks, roost in trees, and are known for their loud, repetitive calls—used both to communicate and to alert the group to danger. While they may seem comical in appearance, guinea fowl are highly adaptive, clever, and protective, embodying the symbolic message that true strength lies in community, vigilance, and staying grounded in who you are.

Gull
(Larus – genus)

spirits of sailors

The gull is a messenger of liminality—a creature that thrives at the threshold between worlds: land and water, sky and earth, wilderness and civilization. In Celtic and Norse traditions, gulls were thought to carry the souls of the drowned, or to be the spirits of sailors who had died at sea. Their cries, often heard before a storm or along lonely coastlines, have long been interpreted as omens, warnings, or calls from the spirit world. In Native American symbolism, particularly in coastal tribes, gulls are seen as clever survivors and scouts, messengers between the seen and unseen, who bring knowledge of approaching change. Spiritually, the gull represents resourcefulness, emotional flexibility, and the ability to find opportunity in chaos. Unlike birds that live in isolation or rigid systems, the gull adapts constantly—moving with the tides, scavenging when needed, and flying when the winds shift. It reminds us to loosen our grip on control, trust in our instincts, and embrace the unknown.

There are over 50 species of gulls, with the herring gull (Larus argentatus), black-headed gull (Chroicocephalus ridibundus), and laughing gull (Leucophaeus atricilla) among the most well-known. Gulls are found on every continent, from icy polar regions to tropical shores, and have proven exceptionally successful in urban, rural, and coastal environments alike. They thrive in temperate to polar climates, often nesting in colonies on rocky islands, dunes, or even rooftops. Gulls are opportunistic omnivores, feeding on fish, refuse, crustaceans, and even other birds' eggs. Their intelligence is well-documented—they use tools, drop shellfish from great heights to crack them, and even mimic other species' calls. Social, vocal, and endlessly adaptive, gulls are survivors and shapeshifters, perfectly suited for the shifting edges of the world. In symbolic terms, they teach us to navigate change with grace, to find sustenance where others see only waste, and to honor the sacred in the in-between spaces.

Harpy Eagle
(Harpia harpyja)

spiritual power

In mythology and symbolism, the harpy eagle is often seen as a guardian of the natural world—a living avatar of the forest's hidden power and spiritual authority. Its name comes from the Harpies of Greek mythology, winged spirits of storm and vengeance, which gives the bird an added aura of mythic ferocity and divine retribution. In Indigenous cultures of Central and South America, the harpy eagle is often revered as a spiritual protector, associated with the sky realm, ancestral wisdom, and the soul's ability to soar beyond earthly limits. To encounter one is to be in the presence of something ancient and sacred. With its intense gaze and nearly supernatural silence in flight, it symbolizes clarity, control, and the balance of immense strength with absolute precision. It is not merely a predator, but a symbol of uncompromising vision and the raw, untamed intelligence of nature itself.

The harpy eagle is native to the rainforests of Central and South America, particularly in countries like Brazil, Panama, Ecuador, and Peru, where it inhabits lowland tropical rainforests with dense canopies. It is one of the largest and most powerful eagles in the world, with females reaching up to 20 pounds (9 kg) and wingspans over 7 feet (2.1 meters). Despite its massive size, it is built for maneuvering through dense foliage, using its incredibly strong talons—the largest of any eagle species—to hunt monkeys, sloths, and large birds. It prefers humid, equatorial climates, nesting high in the tallest trees and returning to the same site over many years. As a top predator and keystone species, its survival is crucial to the health of the rainforest ecosystem. Sadly, due to deforestation and habitat fragmentation, the harpy eagle is now near threatened. In symbolic terms, it reminds us of the fragile majesty of wild places and the need to protect what is both powerful and sacred. The harpy eagle teaches us to see with fierce clarity, to move with strength and silence, and to honor the guardianship of nature itself.

Hawk
(Buteo spp.)

higher awareness

The hawk is a messenger of higher awareness and truth, often associated with intuition, observation, and taking action at the right moment. In Native American traditions, hawks are considered sacred messengers, delivering insight from the spirit world. Their appearance is often interpreted as a sign to pay attention to subtle signs, trust your instincts, and rise above confusion. In Ancient Egypt, the hawk (especially the species associated with the god Horus) symbolized royalty, divine protection, and the eye that sees all. Hawks are also found in Celtic and Norse mythologies as symbols of guardianship, intelligence, and spiritual vigilance. Because hawks are known to fly high while maintaining an eye on even the smallest movement below, they are often used as symbols of long-range vision, higher thinking, and mental precision. They call on us to observe before we act, strike only when necessary, and keep our eyes on what truly matters.

Hawks are found across every continent except Antarctica, with species ranging from the red-tailed hawk (Buteo jamaicensis) in North America to the Eurasian sparrowhawk (Accipiter nisus) and African harrier-hawks. They inhabit a wide variety of environments, including forests, deserts, grasslands, and mountains, and they are adaptable to both temperate and tropical climates. Hawks are birds of prey, feeding on small mammals, reptiles, and other birds, using both soaring and ambush hunting techniques depending on the species. Known for their exceptional eyesight, hawks can spot prey from great distances. They are solitary hunters but often return to the same nesting sites year after year with a lifelong mate. Their silent flight and commanding presence remind us that true power doesn't shout–it sees clearly, waits patiently, and moves only when the time is right. In every tradition, the hawk teaches us to rise above the noise, sharpen our perspective, and walk the path of clarity and quiet strength.

Heron
(Ardea herodias

wisdom

In mythology and spiritual symbolism, the heron is seen as a messenger of stillness and wisdom, often associated with introspection, longevity, and walking one's own path with quiet strength. In Ancient Egyptian mythology, the heron—especially the bennu bird, which closely resembled a heron—was linked to creation, the rising sun, and rebirth, considered a sacred creature connected to the gods and the flow of time. In Chinese symbolism, herons represent purity, strength, and good fortune, while in Native American cultures, particularly among the Pacific Northwest tribes, the heron is seen as a symbol of determination, self-sufficiency, and the wisdom that comes from solitude. Because of its ability to stand motionless for long periods and strike with sudden precision, the heron is also a symbol of mindfulness, timing, and the mastery of restraint—a teacher of doing less to achieve more.

Herons are found across every continent except Antarctica, with species like the great blue heron (Ardea herodias), grey heron (Ardea cinerea), and black-crowned night heron (Nycticorax nycticorax) thriving in wetlands, lakes, rivers, marshes, and coastal shorelines. They favor temperate to tropical climates, often nesting in tall trees or reedbeds and feeding on fish, amphibians, insects, and small mammals. Herons hunt by wading slowly or standing completely still, using their long necks and sharp beaks to strike with lightning speed. Their deliberate movements and solitary nature make them deeply symbolic of the power of silence, the necessity of reflection, and the beauty of self-reliant grace. In both art and spirituality, the heron invites us to slow down, connect with our intuition, and find strength in presence rather than haste.

Hoopoe
(Upupa epops)

spiritual insight

In symbolism and mythology, the hoopoe holds a revered place. In Islamic tradition, the hoopoe is known as the messenger of King Solomon, a creature of intelligence and spiritual insight that helped guide the king with divine information. This story appears in the Qur'an, where the hoopoe discovers the Queen of Sheba and informs Solomon, acting as a diplomatic link between two great leaders. In this role, the hoopoe represents wisdom, communication, and the ability to bridge worlds—between the material and spiritual, or the known and the hidden. In ancient Egyptian and Greek cultures, the hoopoe was associated with protection, transformation, and royal dignity, often seen as a guide between life and death, due to its eerie, otherworldly call. In Persian poetry, the hoopoe appears in The Conference of the Birds by Attar of Nishapur, where it serves as the spiritual leader guiding a group of birds on their quest to find the divine Simurgh. Here, it symbolizes the illuminated soul, the wise teacher, and the seeker who has become the guide.

The hoopoe is native to Europe, Asia, and Africa, favoring open woodlands, orchards, grasslands, and savannahs, especially where old trees and crevices provide nesting spots. It thrives in temperate to tropical climates, migrating in some regions depending on the season. Recognizable by its fan-shaped crest, black-and-white striped wings, cinnamon body, and long, slender beak, the hoopoe uses its bill to probe soil and dung for insects and larvae. It's a solitary nester, often choosing cavities in trees or walls, and can emit a strong musky odor—once thought to ward off predators, which links it further to ancient symbolism around protection, mystery, and ritual cleanliness. Its unusual, elegant appearance and rich mythological history make it a deeply spiritual bird—a symbol of divine insight, sacred duty, and the wisdom that comes from inner clarity.

Hornbill
(Bucerotidae - family)

protection

In Southeast Asian mythology, particularly among Dayak, Iban, and other Indigenous peoples of Borneo, the hornbill is considered a sacred messenger from the spirit world. It is often associated with divine kingship, ancestral spirits, and the heavens, and its feathers are used in ceremonial garb to invoke power, protection, and spiritual alignment. The rhinoceros hornbill, in particular, is seen as the embodiment of a supreme god or spirit, with its distinctive casque echoing both the sun and spiritual authority. Among some cultures, hornbills are symbols of fidelity and strong familial bonds, as their nesting behavior involves the female sealing herself inside a tree cavity, where she is fed by her mate throughout the incubation period—a profound image of trust, unity, and commitment. In African folklore, the hornbill is often seen as a symbol of fertility, renewal, and prophetic vision, sometimes tied to rituals of agricultural abundance and rain-calling.

Hornbills are native to tropical and subtropical Africa and Asia, including India, Thailand, Malaysia, Indonesia, and the Philippines, thriving in dense forests, savannas, and mountainous woodlands. These birds favor warm, humid climates, often living high in the canopy where they feed on fruits, insects, and small animals. They are known for their dramatic beaks and casques—some hollow, others solid—that may be used for sound amplification, sexual display, or headbutting in territorial disputes. Hornbills are exceptional flyers, yet their social and reproductive lives are what make them symbolically rich: their cooperative parenting and strong pair bonds reflect themes of trust, loyalty, and mutual dependence. Sadly, many hornbill species are now threatened by deforestation and hunting, making them not only spiritual icons but also powerful emblems of conservation and the sacredness of living balance. The hornbill teaches us to honor our connections, protect what is precious, and carry our voices boldly through the canopy of life.

Hummingbird
(Trochilidae - family)

messengers of the gods

Across the Americas, where all hummingbird species are native, these birds hold deep spiritual and mythological significance. In Aztec mythology, hummingbirds were connected to the sun god Huitzilopochtli, who was believed to be reincarnated in hummingbird form—making them symbols of warriors, messengers of the gods, and protectors of the spirit. The Aztecs wore hummingbird talismans into battle, seeing them as emblems of vitality, courage, and swift divine energy. In Andean cultures, the hummingbird is seen as a link between the living and the dead, flying effortlessly between worlds and bringing messages of love, remembrance, and hope. In Native American symbolism, particularly among the Hopi and Zuni, hummingbirds are bringers of rain, fertility, and harmony with nature. In modern spiritual teachings, the hummingbird reminds us to live lightly, savor each moment, and find sweetness even in the smallest of things—a totem of joy, movement, and spiritual grace.

There are over 300 species of hummingbirds, ranging from Alaska to the tip of South America, though they are most diverse in Central and South America, especially the Andes. They thrive in tropical and temperate environments, from cloud forests and mountain valleys to deserts and gardens. Despite their size, hummingbirds are fierce and resilient: they migrate thousands of miles, remember every flower they've visited, and have the highest metabolism of any bird. They feed on nectar, insects, and sap, hovering in place or darting with astonishing agility thanks to their ability to beat their wings up to 80 times per second. Their shimmering plumage—often iridescent blues, greens, and reds—makes them look like living jewels, fleeting yet unforgettable. Symbolically, they teach us that lightness, grace, and endurance are not contradictions, but powers that live in harmony. The hummingbird is a master of the moment, a creature that invites us to live from the heart, chase beauty, and sip sweetness wherever we find it.

Ibis
(Threskiornithidae - family)

spiritual order

In Ancient Egyptian mythology, the ibis was sacred to Thoth, the god of wisdom, writing, and magic. Often depicted as a man with the head of an ibis, Thoth was believed to have invented language and maintained the balance of the cosmos through logic and understanding. Because of this connection, the ibis became a symbol of divine intellect, spiritual order, and the pursuit of knowledge. Mummified ibises were offered in massive numbers at temples to honor Thoth, and their presence near the Nile linked them to the life-giving flow of water, cycles, and fertility. In Greek and Roman texts, the ibis was sometimes associated with purity and medicine, believed to possess healing powers. In more modern symbolism, particularly in spiritual and esoteric traditions, the ibis represents discernment, ritual awareness, and the ability to navigate both the material and mystical realms—its curved beak like a key to unlock hidden truths.

Ibises are found on every continent except Antarctica, but are most closely associated with Africa, the Middle East, and the Americas, thriving in wetlands, riverbanks, mangroves, and tidal flats. They prefer warm, humid climates, though some species, like the American white ibis (Eudocimus albus), are common in southeastern parts of the U.S., especially Florida and Louisiana. Ibises feed by sweeping their long bills through shallow water and mud, searching for insects, crustaceans, and small aquatic creatures. Their communal roosting and nesting behavior makes them symbols of harmony, cooperation, and ancestral continuity. With their slow, deliberate steps and elegantly curved form, they embody the energy of ritual grace and spiritual refinement. The ibis teaches us to walk gently through the world, to seek knowledge through experience, and to honor the sacred in daily rhythms and the spaces between things.

Ivory-Billed Woodpecker
(Campephilus principalis)

rare beauty

Symbolically, the ivory-billed woodpecker represents the unknown, the nearly lost, and the power of belief against disappearance. Dubbed the "Lord God Bird" because that's what people supposedly exclaimed upon seeing its magnificent wingspan and bold markings, the ivory-billed holds a special place in the lore of the American South—particularly in the swampy bottomland forests of the southeastern United States and Cuba. In cultural imagination, it has become a ghost bird, a living emblem of what we risk losing when we fail to protect wild spaces. In more mystical terms, it symbolizes rare beauty, the persistence of hope, and the emotional pull of things we can't quite grasp but cannot forget. Its dramatic black-and-white plumage and striking ivory-colored bill give it a totemic presence—representing the unseen power of nature, the tension between extinction and rediscovery, and the deep longing to reconnect with the wild before it's gone.

The ivory-billed woodpecker was once found across southeastern U.S. swamp forests and parts of Cuba, inhabiting dense, mature bottomland hardwood forests rich in deadwood, where it fed primarily on beetle larvae and insects hidden under bark. It required large territories and expansive, untouched forest ecosystems—an ecological need that ultimately led to its downfall as logging and habitat destruction swept through the 19th and 20th centuries. With a body nearly 20 inches long and a wingspan of over 30 inches, it was the largest woodpecker in North America. Males sported a bold red crest, while females bore an elegant black one. Officially declared possibly extinct, reports of sightings persist, and efforts to locate surviving populations have become a symbol of conservation urgency and collective yearning. Whether or not the ivory-billed still glides through the shadows of ancient trees, it lives on as a profound symbol of the delicate edge between memory and miracle, loss and legacy.

Jabiru
(Jabiru mycteria)

quiet guardianship

The jabiru often represents sacred stewardship and balance between earth and sky. In Indigenous South and Central American cultures, the jabiru is viewed as a guardian of wetlands, a spiritual figure that watches over water and life. Its great height, snowy white plumage, and solitary habits give it an aura of solemnity and reverence, marking it as a symbol of patience, awareness, and the stillness required for true observation. In modern spiritual readings, the jabiru is associated with renewal, cleansing, and the importance of protecting delicate ecosystems. Its presence signals a healthy, functioning environment, and its calm, slow movements contrast beautifully with its ability to soar into the sky, representing the power of quiet guardianship. Unlike more vocal birds, the jabiru expresses itself through presence and poise—a reminder that not all wisdom needs to shout.

The jabiru is found in tropical and subtropical regions of the Americas, particularly in Brazil, Paraguay, northern Argentina, Belize, and parts of Mexico, favoring flooded savannas, marshes, riverbanks, and shallow lakes. It thrives in warm, wet climates, often wading through water in search of fish, frogs, mollusks, and carrion. Towering over other birds at up to 5 feet (1.5 meters) tall with a wingspan nearing 9 feet (2.75 meters), the jabiru is unmistakable with its snow-white body, dark featherless head, and distinctive red neck pouch, which it inflates during courtship. It is a monogamous and highly territorial nester, building enormous nests—sometimes reused and added to over several years. Its role as a top wetland dweller and scavenger places it as a key figure in ecosystem balance, and its symbolic resonance reminds us of the importance of preserving peace, standing tall in purpose, and living in harmony with the cycles that sustain us.

Jacana
(Jacanidae - family)

mindfulness

In symbolic terms, the jacana embodies the mastery of delicate balance, both literally and metaphorically. Found in many Indigenous myths and spiritual traditions in South and Central America, Africa, and Southeast Asia, it's often seen as a guardian of wetlands, an emissary between the elements, gliding over the surface where water meets sky and earth. Its movements suggest walking on water, turning it into a symbol of miraculous grace, emotional agility, and the power to remain composed amid shifting tides. In spiritual and psychological symbolism, the jacana also represents the ability to step lightly across emotional challenges—not to avoid them, but to move over them with mindfulness. What makes the jacana especially unique, however, is its reversal of traditional gender roles: females are larger, more territorial, and polyandrous, often mating with several males who do all the incubation and chick-rearing. This has led to its interpretation as a symbol of feminine strength, the divine feminine in leadership, and the power of nurturing through flexibility and role-shifting.

Jacanas are found in tropical and subtropical wetlands across the globe, including the Amazon basin, Central America, sub-Saharan Africa, and Southeast Asia. They thrive in shallow freshwater habitats, where dense vegetation provides both cover and a walking surface. Their incredibly long toes and claws are uniquely adapted to distribute weight across broad, flat surfaces, allowing them to walk across lily pads and floating vegetation. Their diet consists mainly of insects, snails, and seeds picked delicately from the surface. Despite their peaceful appearance, jacanas are fierce defenders of territory, particularly the dominant females, who maintain harems of male partners and patrol large areas. In every way, the jacana defies expectations—teaching us that true strength lies in balance, that roles can be fluid, and that walking with lightness doesn't mean walking without power.

Jackdaw
(Corvus monedula)

lure of mystery

The jackdaw is a bird of perception, curiosity, and hidden wisdom. In Celtic and Norse traditions, jackdaws were seen as watchers and omens, birds that could move between the seen and unseen with ease. Their pale, intelligent eyes and habit of watching from rooftops and ruins gave them a ghostly, mystical aura—often interpreted as messengers between the spirit world and the living. In Greek mythology, jackdaws were sacred to Athena, goddess of wisdom, further reinforcing their association with keen intellect and sharp observation. Because of their habit of collecting shiny objects, jackdaws have also come to symbolize a love of beauty, the lure of mystery, and the instinct to seek value in what others discard. In dream and folklore symbolism, they're seen as creatures that encourage us to pay attention to small signs, think independently, and question the obvious—carrying messages for those willing to look deeper.

The jackdaw is native to Europe, North Africa, and western Asia, and has adapted remarkably well to urban, rural, and forested environments. It thrives in temperate climates, nesting in crevices, chimneys, cliffs, and hollow trees. Jackdaws are highly social birds, often forming lifelong bonds with a mate and staying close to their flock. Their calls are soft but varied, and they communicate constantly through body language and vocalization. As part of the corvid family, jackdaws are extremely intelligent—capable of problem-solving, using tools, and recognizing human faces. Despite their modest size, their behavior and sharp awareness place them among the most intellectually advanced birds in the world. As a totem, the jackdaw invites you to observe rather than react, to value connection, and to embrace your own unique lens on the world. It is a quiet guide in a noisy world—reminding you that insight is found not through noise, but through watchfulness and subtle intuition.

Jay

(Cyanocitta cristata)

voice of the soul

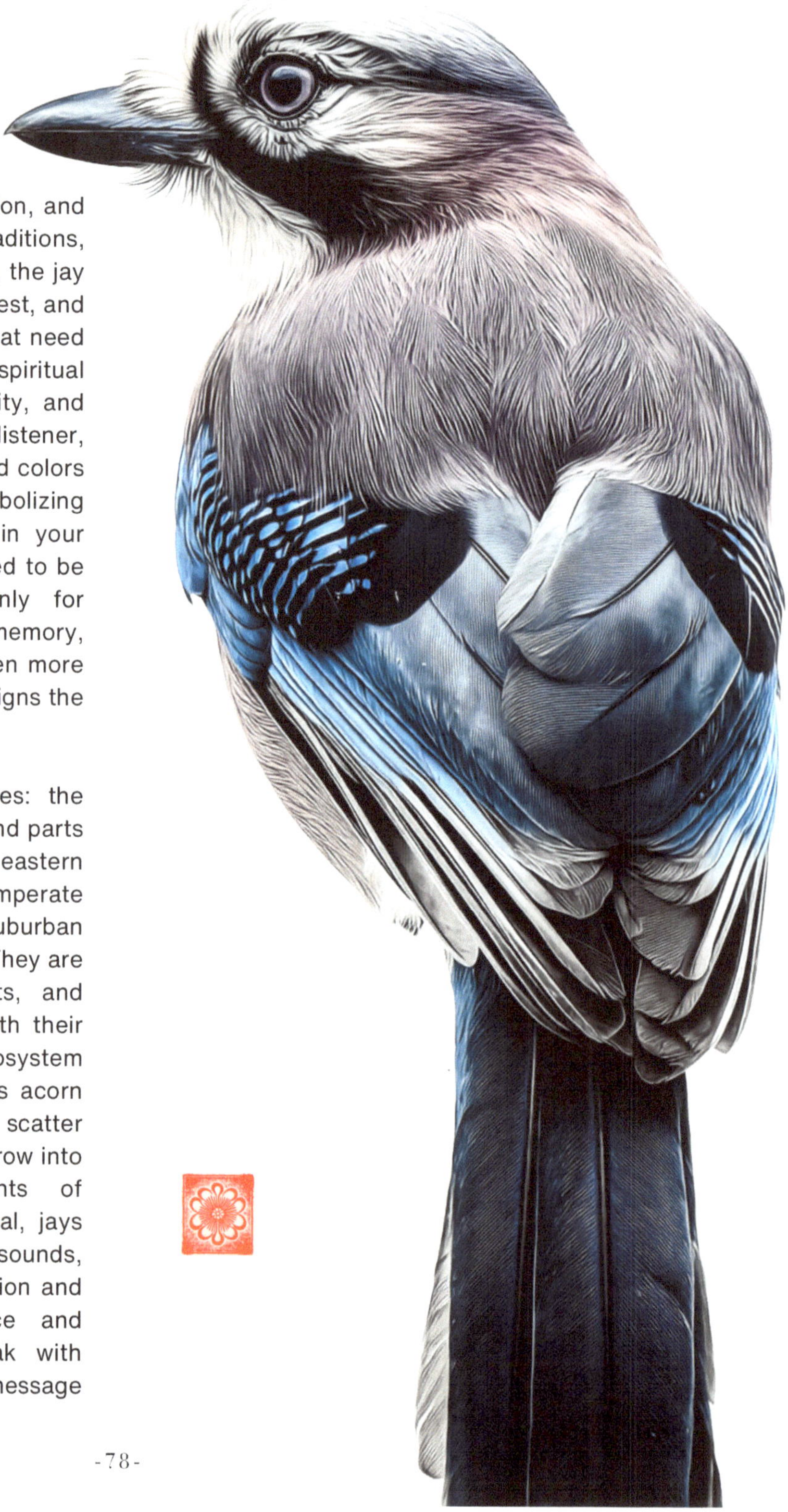

The jay is a messenger of clarity, bold expression, and the voice of the soul. In Native American traditions, especially among the Chinook and Coast Salish, the jay is seen as a talkative spirit, a guardian of the forest, and a trickster figure who calls attention to truths that need to be heard—even if they're uncomfortable. In spiritual symbolism, the jay represents honesty, curiosity, and the ability to adapt one's message to fit the listener, much like a skilled storyteller or teacher. Its vivid colors and loud calls make it impossible to ignore, symbolizing the courage to speak your truth and stand in your individuality. In Celtic folklore, jays were believed to be keepers of secrets, storing acorns not only for sustenance but as symbols of hidden wisdom, memory, and foresight. Their presence was a call to listen more deeply, observe closely, and treasure the small signs the world offers.

Different jay species inhabit varied landscapes: the Eurasian jay is native to Europe, North Africa, and parts of Asia, while the blue jay is found throughout eastern and central North America. Jays thrive in temperate woodlands, mixed forests, and even suburban environments, demonstrating high adaptability. They are omnivorous, feeding on seeds, nuts, insects, and occasionally eggs or nestlings—showcasing both their opportunistic nature and their role in ecosystem balance. One of their most fascinating traits is acorn caching, especially in Eurasian jays, who scatter thousands of acorns each year, many of which grow into oak trees—making them accidental agents of reforestation. Socially complex and highly vocal, jays also mimic other birds and even human-made sounds, underscoring their symbolic ties to communication and intelligence. With their dazzling appearance and assertive presence, jays remind us to speak with conviction, trust in our vision, and carry our message with both clarity and color.

Junco
(Junco hyemalis)

inner peace

In symbolic terms, the junco represents inner peace, humility, and the restorative power of stillness. Because it is most visible during the cold months—arriving with the snow and departing with the spring—many cultures and spiritual interpretations see the junco as a harbinger of change, rest, and quiet reflection before renewal. Its understated grey-and-white plumage evokes themes of purity, modesty, and the hidden beauty in things overlooked. In spiritual readings, the junco often appears as a guide for those seeking clarity or healing, reminding us that growth does not always begin with grand gestures but often with small, quiet choices. The bird's preference for ground-feeding and its simple, rhythmic movements reflect a message of being grounded, honoring your pace, and finding nourishment in life's most elemental forms. It teaches us that there is power in showing up softly, consistently, and without needing applause.

The dark-eyed junco (Junco hyemalis) is the most widespread species, native to North America, and found from Canada and Alaska through the United States and into northern Mexico. These sparrow-sized birds prefer coniferous and mixed woodlands, but are highly adaptable and are often seen in gardens, parks, and rural areas, especially during migration and winter. They thrive in cool to cold climates, particularly in snowy regions, where their soft grey upperparts and white bellies provide camouflage and contrast. Juncos feed primarily on seeds and small insects, foraging in flocks close to the ground. During the breeding season, they nest in hidden, protected spots, often on the forest floor. Though they may seem unassuming, juncos are resilient travelers and seasonal markers—embodying the idea that even in the quietest, coldest seasons of life, there is still movement, grace, and the gentle promise of spring.

Kakapo
(Strigops habroptilus)

survival against the odds

The kākāpō represents gentle strength, survival against the odds, and the wisdom of standing apart. In Māori culture, where the bird is native, it was once admired for its soft feathers and curious personality, appearing in stories as a humble forest dweller with sacred ties to the natural world. Unlike most birds of power or speed, the kākāpō teaches through its stillness, patience, and quirky independence. In a modern context, it has become a global emblem of vulnerability and guardianship—a reminder of what is lost when ecosystems are damaged, but also of what can be saved through dedication and care. Spiritually, the kākāpō symbolizes soulful individuality, slow and meaningful progress, and the power of evolving on your own terms. It's a bird that tells us it's okay to move at your own pace, to be different, and to hold firm in your uniqueness —even in a world moving too fast.

The kākāpō is found only in New Zealand, and is critically endangered, with fewer than 300 individuals alive today, all carefully monitored by conservationists. It once inhabited the dense forests and scrublands of both the North and South Islands, but now survives only on predator-free offshore sanctuaries, like Whenua Hou and Anchor Island. It thrives in cool, temperate rainforests, where it feeds on native plants, fruits, seeds, and leaves. The kākāpō is heavy-bodied and can't fly, but it climbs with strong claws and glides short distances from trees. Males perform a rare, booming lekking call—a low-frequency sound that echoes through valleys to attract females during their irregular, mast-seed-driven breeding seasons. In every way, it is an evolutionary outlier: ancient, herbivorous, slow to mature, and yet charmingly charismatic. As a totem, the kākāpō reminds us to protect what is rare, celebrate our differences, and never underestimate the quiet force of persistence.

Kestrel
(Falco tinnunculus)

divine insight

In mythology and spiritual symbolism, the kestrel has long been associated with keen observation, strategy, and subtle strength. In Ancient Egypt, falcons—including the kestrel—were sacred to Horus, the god of the sky, and considered sunbirds who flew with divine insight and spiritual authority. The kestrel's ability to hover midair became symbolic of pausing before making an important decision, and its sharp eyes and swift strike turned it into a metaphor for clarity, timing, and precision in execution. In Celtic and Norse traditions, the kestrel is connected with freedom of spirit, personal sovereignty, and the ability to rise above distraction. As a totem, it teaches us to remain still, assess wisely, and act swiftly when the moment arrives. Because it is smaller than other raptors yet still a fierce and effective hunter, the kestrel also represents the idea that power is not measured in size, but in clarity of intent.

Kestrels are found across much of the world. The common kestrel (Falco tinnunculus) is widespread in Europe, Asia, and Africa, while the American kestrel (Falco sparverius)—sometimes called the "sparrow hawk"—ranges throughout North and South America. They prefer open habitats such as grasslands, meadows, farmland, and deserts, thriving in temperate to semi-arid climates. One of their most striking behaviors is hovering in place while hunting, known as "wind-hovering," which allows them to scan for prey such as small rodents, insects, and reptiles. Kestrels often nest in tree cavities, cliffs, or even man-made structures, and are known for their adaptability to human-altered landscapes. With their rust-red plumage, bold facial markings, and energetic presence, kestrels are visually and spiritually captivating. They remind us to stay grounded while trusting in our ability to soar—to see clearly from above, and to act with intention, grace, and fearless focus.

Kingfisher
(Alcedinidae - family)

touched by the sun

The kingfisher is a messenger of tranquility, illumination, and the soul's ability to find sustenance in stillness. In Greek mythology, the kingfisher is tied to the legend of Alcyone and Ceyx, a tragic love story that ends in divine compassion as the gods transform the lovers into kingfishers and grant them a period of calm seas to nest—thus creating the term "Halcyon days," a symbol of peace, renewal, and harmony. In Celtic lore, the kingfisher was believed to be touched by the sun, capable of guiding souls between worlds and illuminating unseen truths. In Native American symbolism, the kingfisher is a spirit of prosperity, patience, and timing—often appearing as a totem that urges one to observe before leaping and to dive deep when the moment is right. Its connection to water and air—two symbolic realms of emotion and spirit—makes it a powerful emblem of intuition, spiritual balance, and the courage to pierce through illusion.

Kingfishers are found across Africa, Asia, Europe, Australia, and the Americas, with over 90 species adapted to a variety of habitats, from dense forests and mountain streams to coastal mangroves and desert oases. The most iconic species include the common kingfisher (Alcedo atthis), with its radiant sapphire-and-orange plumage, and the belted kingfisher (Megaceryle alcyon) of North America, known for its rattling call and mohawk-like crest. Kingfishers thrive in temperate to tropical climates, often nesting in riverbanks or termite mounds, where they dig long tunnels for their eggs. Their most remarkable hunting trait is their ability to hover and dive with pinpoint accuracy, catching fish and aquatic insects with incredible precision. They symbolize the art of waiting in stillness, choosing the exact moment to act, and emerging from emotional depths with clarity and purpose. To encounter a kingfisher is to be reminded of the beauty in quiet focus, spiritual alignment, and the power of diving deep into your truth—then rising with light in your beak.

Kite
(Milvus milvus)

celestial guardian

The kite represents clarity of thought, effortless awareness, and silent mastery of perspective. In Hindu mythology, the black kite is associated with the goddess Kali and the god Vishnu, seen as a celestial guardian that removes impurities and protects the righteous. In Ancient Egypt, the kite was linked to the goddess Isis, who took the form of a kite bird while mourning and protecting her slain husband Osiris—making it a potent symbol of grief, protection, resurrection, and spiritual flight. In modern spiritual traditions, the kite often appears as a totem for those who seek to detach from mental clutter, focus their vision, and embrace higher awareness. It teaches the value of circling an issue from above, observing from a calm distance before diving into action—an invitation to approach life with grace, patience, and discernment.

Kites are found across all continents except Antarctica, inhabiting savannas, grasslands, wetlands, and open forests, especially in temperate to tropical climates. They are medium-sized raptors with long, angled wings and forked or rounded tails, allowing them to glide effortlessly for long periods, often seen hovering or drifting in lazy circles as they scan the ground for prey. Their diet includes insects, small mammals, carrion, and reptiles, and they are opportunistic hunters—fierce when needed, but never wasteful. Kites are known for their elegant silence in flight, rarely flapping, and their ability to remain suspended in midair with minimal effort reflects their spiritual symbolism of balance, lightness, and focus without force. They nest in tall trees or cliffs, often returning to the same sites year after year with a lifelong mate. The kite reminds us that true strength can be quiet, and true mastery often lies in the ability to rise above turmoil, observe with calm, and act only when clarity guides the way.

Kiwi

(Apteryx - genus)

self-sacrifice

In Māori mythology, the kiwi is a taonga (treasure), deeply revered as a guardian of the forest floor and the spiritual embodiment of whenua (land). According to Māori legend, the kiwi was once a bird of the sky, but sacrificed its ability to fly to save the forest by eating the insects that threatened its trees—a selfless act that elevated the kiwi to sacred status. As such, the kiwi represents self-sacrifice for the good of the community, devotion to duty, and a grounded spiritual path. It is also seen as a symbol of invisibility and inner strength, teaching us that power doesn't need to be seen or loud to be deeply felt. In modern New Zealand, the kiwi is a national symbol and a term of endearment for the people themselves—a reflection of the bird's humble strength, tenacity, and quiet resilience. It speaks to the spirit of home, heritage, and connection to the land.

Biologically, kiwis are flightless, nocturnal birds found only in New Zealand, with five species, including the North Island brown kiwi (Apteryx mantelli) and the great spotted kiwi (Apteryx haastii). They thrive in temperate forests, scrublands, and even remote mountains, where they forage by night, using their keen sense of smell—rare in birds—to locate worms, insects, and seeds. Uniquely, their nostrils are located at the tip of their long beaks, allowing them to sniff the earth as they probe for food. Kiwis have hair-like feathers, strong legs, and powerful claws for digging. They are monogamous, often forming life-long pairs, and lay some of the largest eggs in proportion to body size of any bird species. Slow-moving and vulnerable, kiwis are fiercely protected by New Zealanders and conservation groups, as habitat loss and introduced predators have made them endangered. As a totem, the kiwi reminds us to honor our roots, walk quietly in strength, and remember that true power often lies in staying grounded and close to what truly matters.

Kondor
(Vultur gryphus)

messenger from heaven

In Incan and Andean cosmology, the kondor is one of the three sacred animals, along with the puma and the serpent, representing the upper world or sky realm (Hanan Pacha). The condor is seen as a messenger between the heavens and the earth, capable of carrying prayers to the gods and guiding souls to the afterlife. Its ability to soar on thermals high above mountain peaks symbolizes spiritual transcendence, divine perspective, and freedom from earthly limitations. Among the Quechua and Aymara peoples, the condor is not only revered but also honored in festivals and traditional dances, seen as a being of great wisdom, protection, and cosmic order. It is also a powerful symbol of endurance and renewal, as it feeds on carrion—transforming death into life by purifying the land and keeping ecosystems in balance. Spiritually, the condor teaches us to rise above the immediate, to see with clarity, and to trust in cycles of transformation.

The Andean condor inhabits mountainous and coastal regions of South America, particularly in Argentina, Chile, Peru, Colombia, and Bolivia, where it prefers open grasslands, alpine zones, and cliffs for nesting. It thrives in cool, arid climates, often near the Andes' snow-capped peaks. With a wingspan of up to 10.5 feet (3.2 meters), it is one of the largest birds capable of flight, using thermals and air currents to soar for hours without flapping. Despite its size, it is not a predator but a scavenger, feeding on large carcasses like deer, llamas, and marine mammals. Condors are monogamous, forming lifelong pairs and nesting in high, isolated ledges where they raise a single chick every two years. Once threatened by habitat loss and persecution, the condor has become a symbol of conservation and cultural pride, especially in Argentina and Chile where it appears on national coats of arms. To see a condor in flight is to witness something timeless—a living embodiment of freedom, dignity, and the sacred duty of keeping life's great cycles in motion.

Lapwing
(Vanellus vanellus)

emotional depth

The lapwing carries rich meanings rooted in folklore, spirituality, and emotional depth. In European folklore, especially across Celtic and English traditions, the lapwing has long been associated with mystery and misdirection, often called the "peewit" after its call. Because of its behavior of feigning injury to distract predators from its nest, the lapwing became a symbol of sacrifice and parental devotion—willing to risk itself to protect its young. This act of diversion, paired with its mournful, wavering cry, led to associations with the soul, grief, and love that endures loss. In Christian allegory, the lapwing was sometimes used to represent those who profess outward piety but stray inwardly, owing to its habit of calling from a distance away from its actual nest —though in modern readings, this trait is interpreted more gently as the art of protection through subtlety and diversion. Spiritually, the lapwing is a guardian figure, calling us to defend what matters most not with aggression, but with cleverness, heart, and presence.

The northern lapwing (Vanellus vanellus) is found throughout Europe and parts of Asia, and winters in North Africa, the Middle East, and South Asia. It favors open farmland, wetlands, meadows, and floodplains, thriving in temperate to cool climates, especially in areas with shallow water and soft soil for probing insects and worms. The lapwing is known for its iridescent green-and-purple plumage, broad rounded wings, and distinctive wispy crest. During the breeding season, males perform elaborate aerial displays—flipping and diving through the air while calling—earning them symbolic connections to courtship, flair, and emotional expression. Though often seen in flocks, lapwings are fiercely protective of their nesting territory, and their ground-level nests make them vulnerable to disturbance. In conservation circles, they are considered indicators of habitat health, and their declining numbers in some regions have made them a focus of protection efforts. The lapwing teaches us that being emotionally open and gentle is not weakness—it is strength in a form the world often overlooks. To guard, to grieve, to love—these are sacred flights of the heart.

Lark
(Alaudidae - family)

touched by divine music

In myth, literature, and spiritual symbolism, the lark is one of the most luminous birds—associated with the first light of day, the awakening of consciousness, and the uncontainable beauty of truth expressed through voice. In Greek mythology, the lark was linked to early risers and those touched by divine music. In Christian allegory, it became a symbol of humility and praise, as it sings while ascending—a metaphor for the soul rising toward heaven in joyful surrender. In Celtic tradition, the lark is a creature of the east wind and sunrise, a bird of clarity, new beginnings, and creative fire. Throughout Western literature, especially in the works of Shakespeare, Shelley, and Wordsworth, the lark is a muse: its song a spontaneous outpouring of joy, artistic freedom, and emotional truth. Spiritually, the lark represents inner radiance and the courage to rise even when unseen, calling us to live authentically and sing our own song regardless of who is listening.

Larks are found across Europe, Asia, North Africa, and North America, inhabiting open grasslands, farmlands, heaths, and tundra. They thrive in temperate to arid climates, nesting on the ground and blending seamlessly with the landscape—plain in color, but celestial in voice. The skylark is especially known for its remarkable song-flight: rising nearly vertically into the air, it hovers and sings for minutes at a time before descending slowly, still singing. This upward burst of melody—often at dawn—has made it a timeless symbol of hope, ascension, and the union of earth and sky. Larks feed on seeds and insects, and their close ties to traditional agriculture have made them both beloved and vulnerable, with populations declining in parts of Europe due to changing farming practices. Still, to hear a lark in song is to feel a sudden, living connection to something ancient, bright, and unburdened. The lark teaches us that even the humblest among us can carry heaven in our voices and rise joyfully, even from the barest field.

Laughingthrush
(Garrulax - genus)

dispeller of bad spirits

The laughingthrush is a bird of harmony, family bonds, and emotional intelligence. In Chinese and Himalayan folklore, it is often celebrated for its bright, musical chatter, which is believed to dispel bad spirits and bring laughter and good luck to those who hear it. The bird's tendency to live and forage in tight-knit groups, constantly vocalizing to one another, has linked it to the ideas of cooperation, mutual support, and the importance of community. In spiritual symbolism, the laughingthrush is seen as a guide who reminds us to speak with kindness, stay close to those we trust, and find joy in everyday companionship. Because of its gregarious nature and tendency to mimic other birds or sounds, it also represents adaptability and emotional attunement—a creature that tunes into the energies around it and responds with empathy and song. The laughingthrush is a messenger of emotional clarity, warmth, and shared joy—reminding us that laughter truly is a form of healing.

Laughingthrushes are native to Asia, with species ranging from the Indian subcontinent to China, Southeast Asia, and the Himalayas, living in temperate to tropical forests, bamboo groves, foothills, and shrubby highlands. They thrive in humid, wooded climates, often staying hidden within dense undergrowth. Species such as the white-crested laughingthrush (Garrulax leucolophus) and the chestnut-crowned laughingthrush (Trochalopteron erythrocephalum) are notable for their bold group behavior and intricate songs. These birds are omnivores, feeding on insects, berries, and seeds, and they typically forage on or near the ground. Laughingthrushes are not just vocal—they are vibrant in spirit, constantly in motion, calling, and interacting. Their movements and music bring life to the forest, embodying the spirit of celebration, storytelling, and the kind of wisdom that grows from shared experience. As a totem, the laughingthrush encourages you to lean into laughter, stay connected, and let your voice rise naturally in moments of joy and truth.

Loon
(Gavia - genus)

dream and vision quest

In Native American mythology, especially among the Algonquian peoples, the loon is considered a sacred spirit messenger, often connected to dreams, vision quests, and creation stories. Some legends describe the loon diving to the bottom of the primordial waters to bring up mud that formed the earth, making it a symbol of genesis, transformation, and deep spiritual labor. Because of its eerie, wailing call—heard echoing across mist-covered lakes at dusk or dawn—the loon has been seen as a voice from the spirit world, or a guide through inner darkness and grief. In shamanic traditions, it represents the ability to navigate the unconscious mind, diving into emotional or psychic depths and returning with insight. The loon is often associated with solitude, fidelity, and spiritual resilience, reminding us that there is wisdom in stillness and power in knowing our own depths.

Biologically, loons are native to North America and northern Eurasia, with the common loon (Gavia immer) being the most well-known species in Canada and the northern United States. They thrive in cool, clear freshwater lakes and boreal forests, nesting on the shorelines and living much of their lives on the water. Loons are expert divers, able to descend over 200 feet in search of fish, using their solid bones—unusual for birds—for ballast. Their black-and-white checkerboard plumage, dagger-like beak, and red eyes give them an intense, almost mystical appearance. Loons are monogamous and return to the same lakes year after year, raising their chicks with quiet devotion. While graceful on water, they are nearly helpless on land and require a long "runway" of water to take off, reinforcing their symbolism of sensitivity, emotional depth, and the importance of the right environment for growth and movement. The loon teaches us to embrace introspection, honor emotional truth, and listen to the soul's call as it echoes across the quiet waters of our lives.

Lorikeet
(Trichoglossus moluccanus)

living fully

In symbolism and spiritual folklore, the lorikeet represents unfiltered joy, vibrant communication, and living fully in the present. In Aboriginal Australian stories, rainbow lorikeets are sometimes seen as messengers of the Dreamtime, linking humans with the lush abundance and sacred color of the land. Because of their brilliant, multicolored plumage—especially in the iconic rainbow lorikeet (Trichoglossus moluccanus)—they're also seen as symbols of beauty through diversity and harmony across contrasts. Spiritually, lorikeets are totems of self-confidence and creativity—not afraid to make noise, be seen, or ask for what they want. Their social nature and endless chatter make them associated with friendship, emotional openness, and childlike enthusiasm. They invite us to release fear of judgment and live with more color, laughter, and light—to sing when we feel like it and connect with others without hesitation.

Lorikeets are found primarily in Australia, New Guinea, and surrounding islands, where they thrive in coastal forests, tropical rainforests, woodland edges, and suburban gardens. They favor warm, humid climates and are especially drawn to flowering trees. Unlike most parrots, lorikeets have brush-tipped tongues, perfectly adapted for feeding on nectar and pollen—making them essential pollinators. They also eat fruit and occasionally insects, and are incredibly agile in the treetops, often seen hanging upside down like acrobats. Lorikeets are highly social and vocal, often flying in noisy, colorful flocks and roosting communally. Their close pair bonds, playful behavior, and vivid coloration make them deeply charismatic. As a symbol, the lorikeet reminds us to celebrate life's sweetness, speak boldly, and live each day as if it were painted in the brightest possible hues.

Lyrebird
(Menura - genus)

song and narrative

In symbolism and folklore, the lyrebird represents illusion, artistic mastery, and the sacred role of the storyteller. Named for the shape of its magnificent tail, which resembles the ancient Greek lyre, the lyrebird is a living symbol of song and narrative. In Aboriginal Australian culture, it is often considered a keeper of knowledge and mimic, entrusted with preserving and replaying the sounds of the forest across generations. Its uncanny ability to imitate not just other birds, but also human-made sounds—chainsaws, camera shutters, car alarms—makes it a metaphor for memory, adaptation, and the thin line between authenticity and performance. Spiritually, it invites us to ask: What voices do we carry? Which stories do we repeat? And what is our own song among the echoes? In art and poetry, the lyrebird is frequently seen as a muse-like presence, capable of conjuring the past into the present through sound alone, and thus is a powerful totem of legacy, remembrance, and the creative force that lives in sound and rhythm.

There are two species: the superb lyrebird (Menura novaehollandiae), which inhabits southeastern Australia, and the Albert's lyrebird (Menura alberti), a rarer species found in a small area of Queensland. They thrive in temperate rainforests and eucalyptus woodlands, especially in cool, damp climates with dense underbrush. Lyrebirds are ground dwellers, feeding on insects and invertebrates, and they are solitary, shy creatures despite their loud voices. The males perform astonishing courtship displays—fanning their ornate tail plumes over their back like a peacock and singing complex, layered compositions that can last for up to an hour. These displays are not only visually mesmerizing but aurally enchanting, incorporating as many as 20 different species' calls—some passed down through generations. Symbolically, the lyrebird teaches us that our voice is an instrument of power, and that through mimicry, creativity, and intentional expression, we have the ability to weave beauty from memory, and shape the world through sound.

Magpie
(Pica pica)

sacred and profane

The magpie holds a particularly complex role. In European folklore, especially in the British Isles, the magpie is linked to omens and superstition—"One for sorrow, two for joy..." goes the well-known rhyme, hinting at its role as a bringer of signs and spiritual messages. It's often seen as a liminal creature, living between the worlds of light and shadow, sacred and profane. In Norse mythology, the magpie was considered sacred to Skadi, the goddess of winter and wilderness, and thought to carry the voices of ancestors or fate. In Chinese symbolism, however, the magpie is viewed far more positively: it is a bird of joy, good luck, and marital happiness—a bringer of good news and one who bridges heaven and earth. Because of its incredible intelligence and tendency to mimic sounds, the magpie also symbolizes communication, mimicry, and the value (and danger) of repeating what we hear. Spiritually, it is a totem of perception, contradiction, and the beauty of paradox—encouraging us to embrace complexity, cleverness, and the full spectrum of emotion.

Magpies are found throughout Europe, Asia, North Africa, and western North America, thriving in a wide variety of environments, from forests and farmlands to suburban parks and urban centers. They do well in temperate climates, and are part of the corvid family, which includes crows, ravens, and jays—all renowned for their intelligence. Magpies have been observed using tools, recognizing themselves in mirrors, and solving complex puzzles, putting them among the smartest animals in the world. They are omnivorous, feeding on insects, seeds, carrion, and even small mammals or birds. Their reputation for stealing shiny objects is somewhat exaggerated, but their curiosity and boldness are real. They are social birds with complex communication and strong family structures. The magpie teaches us that not all beauty is straightforward, that truth can live in contradiction, and that sometimes it is the trickster, not the saint, who holds the deepest wisdom.

Mallard
(Anas platyrhynchos)

peaceful progress

The mallard is a powerful emblem of adaptability, emotional clarity, and peaceful progress. In Native American traditions, ducks like the mallard represent emotional depth, intuition, and the sacred connection to water, often linked to healing, fertility, and feminine wisdom. Their calm appearance on the surface—while paddling furiously underneath—has made them metaphors in both folklore and psychology for grace under pressure and the hidden labor beneath serenity. In Celtic mythology, ducks were seen as shape-shifting creatures, capable of moving between the worlds of land, water, and sky—giving them an almost shamanic role as bridgers of the physical and spiritual realms. The mallard, with its vibrant green-headed male and speckled brown female, is also symbolic of duality and partnership, as it often appears in mated pairs and returns faithfully to nesting sites each year. Spiritually, the mallard reminds us to move with life's current, trust in the wisdom of our emotions, and remain calm while navigating the hidden depths of the self.

The mallard is native to most of the Northern Hemisphere, found throughout North America, Europe, Asia, and even parts of Africa and Australasia due to its success in colonizing urban and rural wetlands. It thrives in temperate climates, frequenting lakes, rivers, ponds, marshes, and city parks, where it feeds on a mix of aquatic plants, insects, crustaceans, and seeds. The male's iconic iridescent green head, white neck ring, and curled tail feathers contrast beautifully with the more camouflaged female, whose mottled plumage is designed for stealthy nesting. Mallards are migratory in colder climates and highly adaptable, often hybridizing with other duck species. They are monogamous for the season, with females taking the lead in nest-building and chick-rearing. Their dabbling feeding style, where they tip forward to graze underwater, adds to their symbolic association with emotional depth, sustenance, and intuitive movement. The mallard teaches us to float gently when the current is calm, paddle when needed, and always find nourishment in what lies just below the surface.

Mandarin Duck
(Aix galericulata)

love and marital devotion

In Chinese, Korean, and Japanese mythology, the mandarin duck is considered the ultimate symbol of lifelong love and marital devotion. Because these birds are known to form strong, monogamous bonds, they are traditionally gifted to newlyweds or featured in art and poetry to represent a harmonious partnership. In Feng Shui, a pair of mandarin ducks placed in the home is said to attract or protect a happy relationship and restore emotional balance. They also symbolize emotional vulnerability and openness, as they are often seen together, floating close on still waters, evoking feelings of trust, gentleness, and deep soul-connection. Their presence in spiritual symbolism reminds us that true love is not just about passion, but about loyal companionship, mutual care, and the strength found in quiet devotion. In modern contexts, they are seen as totems of authentic emotional connection, reminding us to cherish relationships that are both beautiful and lasting.

The mandarin duck is native to East Asia, particularly in China, Korea, Japan, and parts of Russia, though small feral populations now also exist in Europe and parts of California due to ornamental introductions. They thrive in temperate forests near lakes, ponds, and slow-flowing rivers, nesting in tree cavities close to water. Males are especially striking in appearance, with bright orange sails on their backs, iridescent green and purple feathers, a red bill, and striking white markings. Females, more subtly colored in gray-brown with white eye rings, blend beautifully with the forest, ensuring camouflage while nesting. These ducks feed on seeds, aquatic plants, snails, and insects, foraging both on land and in water. Despite their showy appearance, mandarin ducks are surprisingly shy in the wild, preferring quiet, undisturbed areas. Their strong pair bonds, graceful posture, and vibrant harmony in color and movement make them enduring symbols of peaceful union, elegance, and the sacred art of loving without losing oneself.

Martin
(Progne subis)

luck

In folklore and symbolism, the martin is often associated with home, renewal, and divine guardianship. In European traditions, martins—like swallows—were believed to bring good luck to a household, especially if they built nests under eaves or rafters. Destroying a martin's nest was considered extremely bad luck, as these birds were thought to be protected by the Virgin Mary or sacred messengers. In some Christian traditions, martins were linked to the souls of the righteous, or spirits that return each spring as a sign of rebirth and spiritual grace. In Native American symbolism, especially among Southeastern tribes, the purple martin was considered a protector bird, and people would hang gourd houses to invite their presence and ward off pests and negative energy. Spiritually, the martin represents the beauty of freedom with structure—a creature that travels great distances but always returns to its communal roost, reminding us that you can be wildly free and still deeply connected.

Martins are found across North and South America, Europe, Africa, and Asia, with species like the purple martin, sand martin (Riparia riparia), and house martin (Delichon urbicum) being the most well-known. They thrive in temperate and tropical climates, often nesting in colonies under cliffs, bridges, building eaves, or specially made birdhouses. Martins are migratory, traveling vast distances—purple martins, for instance, winter in South America and migrate to North America to breed. They feed on flying insects, catching them midair in dazzling aerial maneuvers. Martins are highly social birds, known for their constant chirping, cooperative nesting habits, and beautiful flight patterns. As a symbol, they encourage us to move freely, speak joyfully, and find beauty in shared experience, reminding us that community is not a cage but a circle of belonging, and that grace lives in both voice and motion.

Merlin Falcon
(Falco columbarius)

mental sharpness

The Merlin falcon represents tactical brilliance, clear vision, and fearlessness in the face of adversity. Once known in falconry as the "lady's hawk" due to its size and elegance, the Merlin has long been associated with nobility, speed, and mental sharpness. Its name, while purely coincidental in origin, resonates with the legendary wizard Merlin of Arthurian myth, adding layers of magic, foresight, and transformation to its symbolism in modern interpretations. Like the mythical Merlin, the bird evokes inner clarity and decisive action, suggesting a totem of quick thinking, adaptability, and strategic insight. Spiritually, it is a symbol of those who do not seek the spotlight, but act with swift precision when called upon, embodying the idea that power does not require grandeur to be effective. The Merlin teaches us to hone our senses, know when to strike, and master the skies of our own potential.

The Merlin is found across North America, Europe, and Asia, thriving in a variety of habitats including open grasslands, coastal cliffs, boreal forests, and tundra, and increasingly adapting to urban settings. It prefers cool to temperate climates, especially during breeding season, but migrates southward in winter. Merlins are small falcons—about the size of a pigeon—but they are tenacious hunters, known for their relentless pursuit of small birds in flight, often using surprise and bursts of speed rather than prolonged soaring. They feed primarily on songbirds, insects, and small mammals, and are often seen flying fast and low in pursuit of prey. With their sleek build, barred plumage, and intense gaze, Merlins are a marvel of compact power and instinct. As a spirit guide, the Merlin falcon calls on us to act without hesitation, fly straight toward our goals, and trust in our finely honed abilities—even when others underestimate us.

Mockingbird
(Mimus polyglottos)

innocence

The mockingbird stands for the purity of voice, the sacredness of innocence, and the act of giving without harming. In Native American traditions, especially among the Cherokee, the mockingbird is a teacher of speech, memory, and spiritual expression, believed to help individuals find their true voice and speak with authenticity. Its most widely known symbolic role comes from Harper Lee's novel To Kill a Mockingbird, where it serves as a metaphor for innocence and moral integrity—"it's a sin to kill a mockingbird" because it does no harm, offering only beauty through song. In this sense, the bird has come to represent those who are vulnerable, generous, and unjustly persecuted. Spiritually, the mockingbird invites us to explore identity through many perspectives, to empathize, to listen deeply, and to express what needs to be spoken, even when the message is borrowed or echoes the pain of others. It is a totem of storytellers, musicians, and truth-bearers.

The northern mockingbird is found throughout North and Central America, from Canada to Mexico and the Caribbean, thriving in open areas with shrubs, lawns, parks, and forest edges. It prefers temperate to subtropical climates and has adapted well to urban environments. What sets it apart is its extraordinary vocal range—a single mockingbird may sing hundreds of different phrases, mimicking other bird species, mechanical sounds, and even human voices. Males sing to attract mates and defend territory, often continuing their performance well into the night. These birds are also fiercely protective of their nests and young, diving at intruders far larger than themselves—adding to their symbolic image of the gentle guardian who will fiercely defend the vulnerable. As a guide, the mockingbird reminds us to use our voice with care and purpose, to draw from experience, and to uplift others through compassion and courage.

Moorhen
(Gallinula chloropus)

omen of transition

The moorhen is closely tied to the element of water, and as such, represents emotional sensitivity, depth, and the ability to move fluidly between worlds—earth and water, visible and unseen. In Celtic and British folklore, moorhens were considered secretive guardians of liminal spaces like marshes and ponds, often appearing just before changes in weather or emotional states, making them omens of transition, intuition, and quiet transformation. Their somewhat awkward, duck-like appearance and solitary yet determined behavior make them symbolic of those who may appear unassuming but are incredibly resourceful and quietly resilient. Because they are fiercely protective of their young and capable of adapting quickly to new environments, they are also symbols of maternal care, boundary-setting, and emotional self-protection. Spiritually, the moorhen reminds us that we don't have to be bold or loud to be anchored, perceptive, and deeply attuned to the flow of life's emotional currents.

The moorhen is found across Europe, Asia, Africa, and the Americas, favoring wetlands, marshes, ponds, rivers, and reed-filled lakes. It thrives in temperate to tropical climates, nesting close to the water's edge in dense vegetation. Despite its duck-like behavior, it is more closely related to coots and rails. The moorhen is recognizable by its dark plumage, bright red and yellow beak, long legs, and white streaks along its sides, and it often walks with a jerky, bobbing motion. It feeds on aquatic plants, small invertebrates, and seeds, often dabbling or picking through waterweed. Moorhens are strong swimmers but also quick on land, and are known for their aggressive defense of territory during breeding season. They nest low to the ground, sometimes building multiple nests for decoys or reuse. As a symbolic presence, the moorhen invites us to stay emotionally grounded in murky times, protect what is ours with quiet determination, and move through emotional or spiritual transitions with grace and intention.

Motmot
(Momotidae - family)

sacred messenger

In symbolism and indigenous mythology, particularly among Mayan and Aztec cultures, the motmot was revered as a sacred messenger of time and cycles. Its rhythmic tail-swaying motion, almost like a ticking clock, gave rise to stories that linked it to the passage of time, destiny, and the balance between worlds. Some myths suggest that the motmot once wore bright feathers as a royal bird, but after failing to show up in battle, it hid in the shadows, only to emerge as a symbol of contemplation and hidden wisdom. Because it sits quietly for long periods before striking with precision to catch prey, the motmot is seen as a teacher of patience, timing, and quiet confidence—reminding us to observe before acting, to trust instinct over impulse, and to let intuition guide the swing of our choices. Its coloration—vivid blues, greens, and oranges—adds to its mystique, linking it to spiritual vision, depth of emotion, and inner clarity.

Motmots are found in tropical and subtropical forests from Mexico through Central America and down into northern South America, especially in Belize, Costa Rica, Colombia, and Brazil. They prefer dense rainforest edges, riverbanks, and shaded jungle paths, where they perch silently on low branches and forage for insects, small reptiles, and fruit. The most distinctive feature of many motmot species, such as the blue-crowned motmot (Momotus coeruliceps) or the turquoise-browed motmot (Eumomota superciliosa), is their racquet-shaped tail feathers—created when the central shaft loses barbs near the tip, creating a unique "swinging pendulum" look. They nest in burrows dug into earthen banks, a trait rare among birds of their visual splendor. As totems, motmots carry messages of quiet brilliance, measured movement, and the wisdom found in stillness. They teach us that life is not always about rushing ahead, but about pausing, perceiving, and acting only when the moment is right.

Myna
(Acridotheres tristis)

marital harmony

In spiritual and folkloric symbolism, the myna is associated with communication, adaptability, and truth spoken through many voices. In Hindu mythology, the myna is sometimes seen as a sacred messenger, associated with love and devotion, particularly when seen in mated pairs, which are believed to bring marital harmony. In Buddhist traditions, the myna is revered as a bird of eloquence, able to recite prayers and sacred chants, and is sometimes seen as a reincarnated monk or teacher. Across South and Southeast Asia, mynas are admired for their mimicry of human speech, which makes them both symbolic and literal mirrors of the human voice—birds that can teach, reflect, or challenge. Their symbolism also touches on the duality of speech: the ability to heal and to deceive, making the myna a totem of discernment, self-expression, and the responsibility that comes with communication.

Mynas are native to South and Southeast Asia, though several species, such as the common myna (Acridotheres tristis), have been introduced to parts of Africa, Australia, the Middle East, and the Pacific Islands due to their adaptability and insect-eating habits. They thrive in tropical and subtropical climates, especially in urban areas, agricultural fields, temple courtyards, and open woodlands. Mynas are omnivorous, feeding on insects, fruits, grains, and scraps, and are known for their fearlessness around humans. They are highly social and often mate for life, nesting in tree cavities, buildings, or even street signs. Their voices are complex and variable—mynas can mimic car alarms, human speech, and the calls of other birds, making them standout members of the starling family. Symbolically, the myna reminds us to listen carefully, speak intentionally, and understand the weight of words. It's a bird of adaptation through voice, teaching us that what we say shapes the world we live in—and that language is a bridge, not just a tool.

Nighthawk
(Chordeiles minor)

spiritual vision

The nighthawk is a creature of the threshold, connected to dusk, dreams, and unseen truths. In Native American traditions, particularly among the Plains tribes, nighthawks were seen as protectors of the night and guardians of sleep, often associated with dreaming, intuition, and spiritual vision. Their erratic, bat-like flight and habit of appearing just as the day ends earned them a reputation as messengers between light and darkness—those who move through the spaces most of us avoid. The distinctive booming dive of the male during mating season adds to their aura of ritual and unseen power. In modern spiritual readings, the nighthawk is often associated with discernment, patience, and the wisdom to navigate the in-between spaces of life—twilight moments, emotional transitions, or creative lulls. As a totem, the nighthawk teaches us that some things are best seen not in the full glare of daylight, but in the softened light of dusk, where shadows tell stories the sun cannot.

The common nighthawk (Chordeiles minor) is native to North and South America, ranging across Canada, the United States, Mexico, and parts of Central and South America, depending on the season. They thrive in open woodlands, grasslands, deserts, and urban areas, often nesting on bare ground, gravel rooftops, or open forest floors. Nighthawks are migratory, flying vast distances—some species migrate over 10,000 kilometers between breeding and wintering grounds. They feed entirely on flying insects, capturing them midair with their wide, gaping mouths, and are most active at dawn and dusk. Camouflaged in mottled grey, brown, and black, they are nearly invisible during the day, blending perfectly into their surroundings. The nighthawk's low, pulsing call and swooping silhouette at dusk make it feel like a ghost moving through the edge of the visible world. As a symbol, it reminds us to trust our instincts in the dark, embrace change, and find peace in the silence between moments.

Nightingale
(Luscinia megarhynchos)

beauty and poetry

The nightingale has long been associated with the soul's song—the idea that beauty, pain, and truth can pour from the heart in the form of voice. In Greek mythology, the nightingale was linked to the tragic figure of Philomela, who was transformed into the bird after suffering great sorrow, her song forever echoing her untold story. This gave rise to the nightingale as a symbol of unspoken emotion, the voice of the silenced, and transformation through suffering. In Persian poetry and Sufi mysticism, the nightingale (or bulbul) represents the lover, crying out to the divine rose, a metaphor for the soul's yearning for union with the divine beloved. In Western Romantic literature, especially through poets like Keats, Shelley, and Coleridge, the nightingale is the ultimate muse—a creature of pure beauty whose song transcends time and mortality, representing the power of art to endure and elevate the spirit. It is a totem of sensitivity, emotional bravery, and finding your voice in the quiet hours of your life.

The nightingale is native to Europe, parts of Asia, and North Africa, thriving in woodlands, scrublands, and riverbanks, especially where dense underbrush provides shelter. It prefers temperate climates, and migrates between southern Europe/northern Africa and central/eastern Europe during the year. Though small and plain in appearance—brownish with no extravagant markings—the male nightingale possesses one of the most celebrated songs in the animal kingdom: a rich, layered, and deeply emotional composition that can last through the entire night. Nightingales are shy and usually hidden, singing from within thick foliage. Their song includes flute-like tones, trills, whistles, and soft phrases, all performed with precision and soul. As a symbol, the nightingale reminds us that great beauty can come from stillness, and that the heart's truth is often most clearly heard when the world grows quiet. It teaches us to honor our feelings, to give voice to longing and love, and to trust that the most meaningful songs are the ones sung without needing to be seen.

Nutcracker
(Nucifraga caryocatactes)

resourcefulness

The nutcracker represents planning, foresight, and mutual dependence with nature. These birds are famous for their ability to harvest and hide thousands of pine seeds, caching them across vast territories to feed on through harsh winters. But they don't retrieve them all—and in doing so, they play a vital role in reforesting the mountains, giving rise to entire generations of pine trees. In Native American and Indigenous lore of the western U.S. and Canada, the nutcracker is sometimes seen as a tree-bringer, a totem of rebirth and resourcefulness, and a guardian of sacred knowledge, especially in high-altitude environments. Spiritually, the nutcracker is a reminder that our efforts today, even if we don't see the results, can shape the world for generations to come. It teaches trust in the cycle of giving and receiving, and the value of quiet, intentional preparation—a bird that works not from fear, but from instinctive wisdom.

The Clark's nutcracker is found in the mountainous western United States and Canada, especially in areas dominated by pine species like whitebark, limber, and ponderosa pine. The spotted nutcracker, its Eurasian counterpart, ranges across northern and central Europe through Siberia and the Himalayas. Both species thrive in cool, high-altitude conifer forests, often nesting in sheltered areas or tree cavities. Their specialized beaks are perfect for cracking tough cones, and their spatial memory is extraordinary—they can remember the locations of thousands of seed caches even under feet of snow. Though solitary during much of the year, they form strong pair bonds during nesting season and are highly protective of their young. As symbolic guides, nutcrackers offer a message of deliberate action, ecological harmony, and legacy-building. They remind us to invest in the future, trust our instincts, and honor the invisible threads between effort and outcome—even if we may never witness the trees that grow from the seeds we've planted.

Osprey
(Pandion haliaetus)

abundance

The osprey represents focus, strategy, and the harmony between opposing elements—air and water, light and depth, thought and action. In Native American traditions, especially among coastal and river-dwelling tribes, the osprey is often seen as a guardian of waterways and a symbol of abundance and divine provision, believed to bring messages from the water spirits or sky gods. Its unmatched ability to dive into water from great heights and emerge with prey links it to the idea of trusting one's intuition and striking at exactly the right moment. In Celtic lore, it is associated with keen perception and purity of intention, admired for its independence and its fierce commitment to survival. The osprey is also seen as a symbol of mastery, often representing those who have cultivated sharp instincts, unwavering will, and a soul-level purpose. Spiritually, it is a totem of self-reliance, elemental balance, and sacred timing—calling us to know when to soar, and when to dive.

Ospreys are found on every continent except Antarctica, thriving near rivers, lakes, marshes, estuaries, and coastal shorelines, wherever there is abundant fish. They prefer temperate to tropical climates, nesting in high, open places—such as trees, cliffs, or even man-made platforms—and are known to return to the same nest year after year, often enlarging it with each season. With their hooked beaks, striking white underbellies, and dark eye stripes, ospreys are visually dramatic. Their diet consists almost entirely of live fish, which they catch by hovering above water and plunging feet-first, using their curved talons and spiny footpads to grip slippery prey. Unlike many birds of prey, ospreys are monogamous and highly cooperative with their mates, both in hunting and nesting. As a symbolic force, the osprey teaches us to trust our instincts, act with decisiveness, and embrace the space where clarity meets courage. It is a reminder that true power lies not in domination, but in focus, patience, and knowing when to strike with precision and grace.

Ostrich
(Struthio camelus)

purity of the soul

The ostrich represents truth, resilience, and adaptability, with deep roots in African, Middle Eastern, and ancient Egyptian traditions. In Egyptian mythology, the ostrich feather was sacred to Ma'at, the goddess of truth and justice—her feather was placed on the scales in the afterlife to weigh the purity of the soul. This made the ostrich a symbol of moral clarity, divine balance, and spiritual integrity. In African folklore, the ostrich is seen as a creature of both wisdom and caution, a teacher of survival in harsh climates, and sometimes a figure of comic irony—mocked for its size or perceived oddities but secretly possessing deep awareness and ancient knowledge. The myth that ostriches "bury their heads in the sand" when afraid is false, yet symbolically it reflects the idea of selective perception, asking us to choose wisely what we focus on, and when it is necessary to face or retreat from conflict. Spiritually, the ostrich invites us to stand tall, trust our instincts, and protect our energy fiercely when needed—knowing that we don't need wings to soar in our own way.

The ostrich is native to Africa, especially in the savannas, deserts, and open woodlands of southern and eastern regions, though it once roamed more widely across the Middle East and North Africa. It thrives in arid to semi-arid climates, and is perfectly adapted to harsh terrain with its long, powerful legs, large eyes, and fast-paced stride, reaching speeds of up to 45 mph (72 km/h). Standing up to 9 feet tall and weighing over 300 pounds, ostriches are unmistakable in form. They are herbivorous, feeding on seeds, shrubs, and insects, and are famous for their protective behavior toward their eggs, which are the largest of any bird species. The males are especially bold, often displaying dramatic dances and booming calls during courtship. In symbolic terms, the ostrich reminds us to embrace what makes us different, move forward with purpose even when we're grounded, and protect our boundaries with quiet confidence and strength.

Owl
(Strigidae / Tytonidae - families)

insight

The owl represents inner vision, ancestral knowledge, and the ability to see through illusion. In Greek mythology, the owl was sacred to Athena, goddess of wisdom and war strategy. The "owl of Athena" became an enduring emblem of intellect, insight, and strategic clarity—not just of knowing, but of knowing when and how to act. In Native American traditions, owls carry layered meanings: among some tribes, they are guardians of sacred knowledge and death, while among others, they are protectors, dream guides, or keepers of truth. Their nocturnal nature makes them symbols of the unconscious mind, and their ability to move silently and strike without being seen imbues them with a sense of mystic precision, spiritual stealth, and otherworldly awareness. In Hindu and Japanese belief systems, owls are linked to wealth and luck, especially as companions of Lakshmi, the goddess of fortune. Overall, owls are powerful archetypes of shadow work, truth-seeking, and transformation through introspection—reminding us that to see in the dark is not to fear it, but to master it.

Owls are found on every continent except Antarctica, inhabiting environments from tropical forests and Arctic tundra to deserts, cities, and grasslands. There are over 200 species, ranging from the tiny elf owl (Micrathene whitneyi) to the imposing Eurasian eagle-owl (Bubo bubo) and the iconic barn owl (Tyto alba). They thrive in nocturnal and crepuscular conditions, thanks to their exceptional vision, acute hearing, and near-silent flight. With their forward-facing eyes, facial discs that direct sound, and flexible necks, owls are built for perception. They feed primarily on small mammals, birds, and insects, often swallowing prey whole and regurgitating the indigestible parts in pellets. Many species are solitary and elusive, but some are more adaptable, nesting in barns, ruins, or forests. As symbols, owls teach us to embrace quiet power, observe with depth, and trust the truths that reveal themselves in silence and stillness. They remind us that not all wisdom is loud, and that the most profound guidance often comes in the darkness, when we learn to truly listen.

Parakeet
(Melopsittacus undulatus)

romantic love

The parakeet represents the heart's voice, the power of words, and the importance of playfulness and connection. In spiritual and cultural symbolism, particularly in Indian and Southeast Asian traditions, parakeets have been associated with romantic love and courtship—often seen perched near lovers in classical paintings or fluttering in poetic imagery, symbolizing the sweetness of expression, flirtation, and shared affection. In Hindu mythology, the green parakeet is the vahana (vehicle) of Kamadeva, the god of love and desire, making it a symbol of attraction, beauty, and the flowering of the heart. Parakeets, like other parrots, are known for their ability to mimic human speech, which connects them to the power of words, storytelling, and the energy we transmit through what we say. As spiritual guides, parakeets invite us to speak with love, surround ourselves with uplifting voices, and find joy in daily connections, no matter how small.

Parakeets are native to Australia, South Asia, and parts of Africa and South America, with different species adapted to savannahs, grasslands, scrub forests, and tropical environments. The budgerigar, or budgie, is native to the Australian interior, where it lives in dry open plains, feeding on seeds and grasses, and moving in flocks that can number in the thousands. Budgies are known for their bright plumage, which ranges from green and yellow in the wild to an astonishing variety of blues, whites, and purples in domestic breeds. They thrive in temperate to warm climates, often forming strong social bonds, learning human words, and showing remarkable intelligence despite their small size. Parakeets are gregarious, curious, and vocal, making them symbols of joyous community, authentic communication, and living life in full color. As spirit animals, they teach us to sing our truth, find beauty in connection, and bring brightness into even the dullest corners of our daily lives.

Parrot
(Psittaciformes - order)

desire and passion

The parrot embodies expression, memory, truth, and the importance of how we use our voice. In Hindu mythology, parrots are linked to Kamadeva, the god of love, and are associated with desire, passion, and the sweetness of poetic speech. In Amazonian tribal cultures, macaws and other parrots are considered sacred birds—mediators between humans and spirits, their plumage used in ceremonial dress to connect with divine forces. In Buddhist tradition, parrots are guardians of sacred teachings, believed to repeat mantras and preserve wisdom, even across lifetimes. Their ability to mimic human speech adds to their symbolism as mirrors of the soul, reflectors of our words, and reminders of the power of intention in everything we say. Some interpretations also see parrots as truth-tellers, unconsciously repeating what has been hidden, revealing what lies beneath the surface. Spiritually, parrots invite us to speak wisely, live boldly, and honor the stories we tell ourselves and others.

Parrots are native to Central and South America, Africa, Asia, and Australasia, inhabiting a wide range of environments including rainforests, savannahs, mangroves, and dry woodlands. They thrive in tropical and subtropical climates, with many species—like the blue-and-yellow macaw (Ara ararauna), African grey (Psittacus erithacus), and ring-necked parakeet (Psittacula krameri)—adapted to life both in the wild and in human company. Parrots are omnivores, feeding on fruit, seeds, nuts, nectar, and sometimes insects. They are renowned for their vocal range, intelligence, social complexity, and strong pair bonding. The African grey, in particular, is famous for its advanced cognitive abilities and emotional sensitivity. As a symbolic guide, the parrot teaches us to celebrate our individuality, share our inner truth with confidence, and fill the world around us with color, clarity, and joy. It reminds us that words are magic, and with them, we build or destroy the stories of our lives.

Peacock
(Pavo cristatus)

knowledge

In Hindu mythology, the peacock is sacred to Saraswati, goddess of wisdom and the arts, and Kartikeya, god of war, who rides the peacock into battle as a symbol of noble power, knowledge, and fearlessness. In Buddhism, it represents openness of spirit, as the peacock can ingest poisonous plants without harm—symbolizing the transformation of suffering into beauty and wisdom. In ancient Persia, it was linked to the divine guardian Simurgh, while in Christianity, especially during the early Byzantine period, it became a symbol of resurrection and eternal life, since it was believed that the peacock's flesh did not decay after death. The "eyes" on its tail were associated with watchfulness, all-seeing protection, and divine omniscience. In a modern spiritual context, the peacock is a totem of authenticity, self-expression, and living in full, unapologetic beauty. It teaches us that true confidence comes not from arrogance, but from embodying our wholeness without shame.

The Indian or blue peafowl (Pavo cristatus) is native to the Indian subcontinent, thriving in forests, scrublands, and temple gardens across India, Sri Lanka, Nepal, and Pakistan. They prefer warm, tropical to subtropical climates and are highly adaptable to both wild and cultivated environments. Males are famed for their brilliant blue heads and necks, long tail feathers with "eye" markings that they fan out in courtship displays, and their piercing, echoing calls. Peacocks are omnivores, feeding on seeds, insects, flowers, and small reptiles. Despite their visual opulence, they are highly alert and quick-footed, often seen as guardians against danger in rural areas due to their keen senses. Peafowl often live in loose flocks and form strong mating bonds. As symbolic creatures, they remind us to walk with grace, reveal our beauty without fear, and protect what is sacred within us. The peacock's dance is not just for show—it is a spiritual invitation to embrace the full spectrum of who we are, with all our colors shining.

Pelican
(Pelecanus - genus)

resurrection

In Christian symbolism, particularly in medieval and Renaissance art, the pelican was believed to pierce its own breast to feed its young with its blood during times of famine. Though biologically inaccurate, this powerful image transformed the pelican into a sacred emblem of Christ-like sacrifice, parental devotion, and unconditional love. It became a metaphor for the Eucharist, and by extension, a symbol of divine mercy and resurrection. In ancient Egypt, the pelican was seen as a protective figure, often featured in funerary texts as a guardian of the dead, associated with rebirth and safe passage into the afterlife. Among coastal Indigenous cultures, the pelican is a totem of abundance and generosity, often linked to the idea of community sharing and the sacred exchange of giving and receiving. Spiritually, the pelican calls on us to care for others without depleting ourselves, to offer from the heart, and to trust that nourishment—emotional or material—flows when it is shared freely.

Pelicans are found across every continent except Antarctica, especially in coastal and inland waterways, where they thrive in temperate to tropical climates. There are eight species worldwide, including the majestic American white pelican (Pelecanus erythrorhynchos) and the brown pelican (Pelecanus occidentalis), the latter being iconic along the shores of the Americas. They feed primarily on fish, using their famous throat pouches to scoop up prey and drain out water. Some species hunt cooperatively, herding fish into shallow waters, a behavior that reflects their symbolic ties to cooperation and communal care. Pelicans are also exceptional fliers—despite their size, they soar gracefully on thermals, gliding effortlessly in formation. They nest in large colonies, often on islands or secluded coasts, raising their young in safe, communal settings. As a symbol, the pelican reminds us to lead with compassion, give without fear, and recognize that true strength lies not just in flight, but in care, commitment, and the quiet rituals of love.

Penguin
(Spheniscidae - family)

spirit of the sea

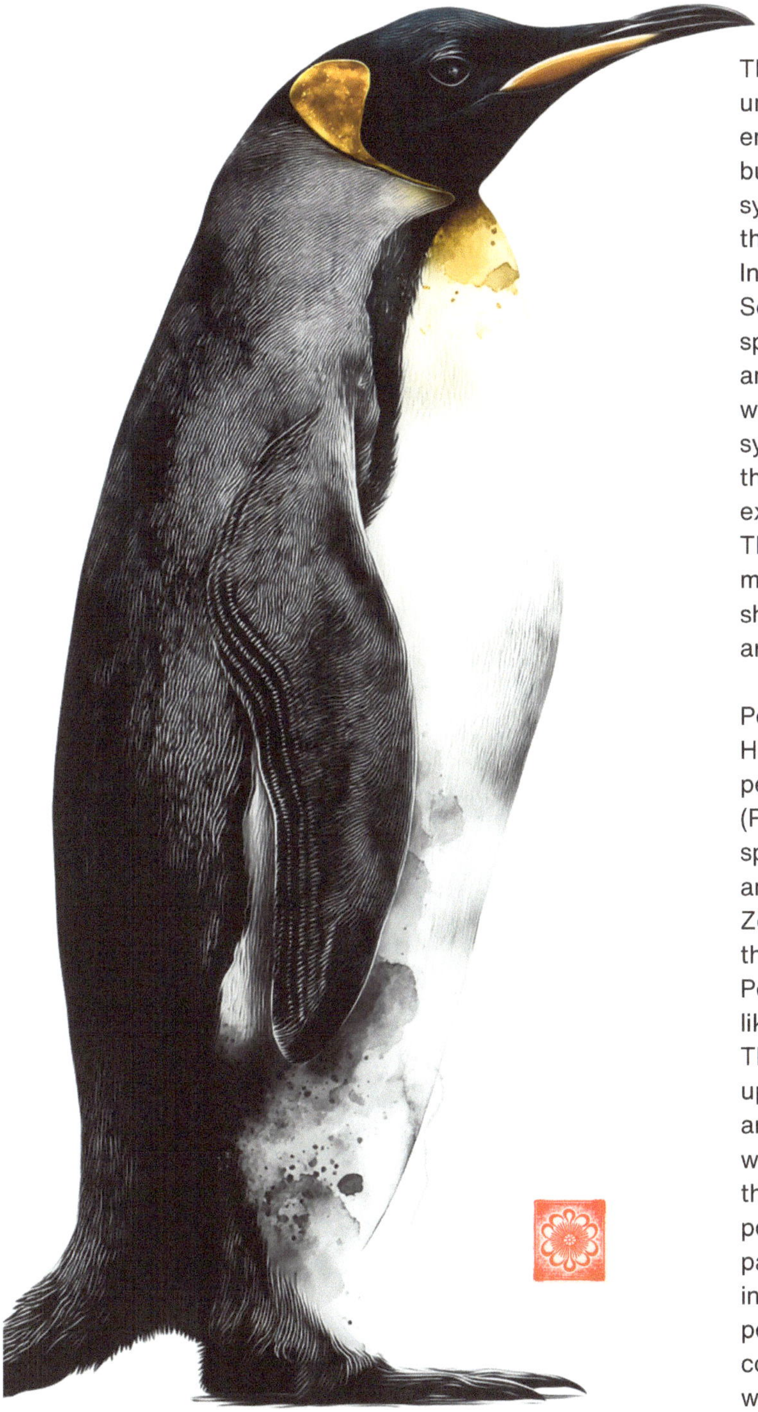

The penguin represents balance, perseverance, and unity, especially in challenging or emotionally "frozen" environments. Because penguins cannot fly through air but soar through water, they are often viewed as symbols of alternative paths to freedom—reminding us that there is more than one way to move forward in life. In Indigenous folklore from the southern regions of South America, the penguin is sometimes seen as a spirit of the sea, a bridge between the living and the ancestral, guiding souls through stormy emotional waters. In modern spiritual interpretations, penguins symbolize devotion and cooperation, particularly in how they take turns incubating eggs, survive through extreme cold, and huddle together to preserve warmth. Their tuxedo-like appearance has also made them metaphors for grace under pressure, encouraging us to show calm, strength, and kindness even when the world around us is harsh and unpredictable.

Penguins are found exclusively in the Southern Hemisphere, with species ranging from the emperor penguin (Aptenodytes forsteri) and Adélie penguin (Pygoscelis adeliae) of Antarctica, to more temperate species like the Humboldt penguin of Chile and Peru, and the little blue penguin of Australia and New Zealand. They thrive in cold marine environments, though some species live in warmer coastal zones. Penguins are excellent swimmers, using their flipper-like wings to "fly" through water at impressive speeds. They feed on fish, squid, and krill, and are often preyed upon by seals and orcas. Most species are highly social and monogamous, raising chicks in large colonies where they cooperate, communicate, and survive through coordination and instinct. The emperor penguin, in particular, is iconic for its incredible parenting rituals, with males incubating eggs for weeks in the Antarctic winter while females hunt. As totems, penguins remind us that true courage can be quiet, that community can save us in the coldest times, and that when we can't fly, we can still glide forward with grace beneath the surface.

Pheasant
(Phasianus colchicus)

prosperity

In Asian and European symbolism, the pheasant has long been associated with nobility, the cycles of nature, and fertility. In Chinese mythology, it is a sacred bird linked to the sun, light, and yang energy, and was often associated with phoenix lore, representing high status, moral virtue, and creativity. The golden pheasant, with its radiant golden crest and crimson body, is still revered as a symbol of royal beauty and spiritual awakening. In Japanese tradition, the pheasant (kiji) is considered the messenger of the gods, especially Amaterasu, the sun goddess, and is also a national bird—symbolizing courage, protection, and vigilance. In European folklore, especially in hunting culture, the pheasant represents the temptations of beauty, as well as cleverness, escape, and wild wisdom—always visible in a flash of color, then gone again. Spiritually, the pheasant teaches us to recognize when to show ourselves and when to retreat, when to express boldly and when to blend into the background. It is a totem of strategic beauty and mindful desire.

The common pheasant (Phasianus colchicus), originally native to Asia, was introduced widely across Europe and North America, where it now thrives in grasslands, forest edges, agricultural fields, and hedgerows. It prefers temperate climates, nesting on the ground and often remaining elusive in dense vegetation. The male is famous for his iridescent plumage—bronze, green, copper, and gold—with a white neck ring and sweeping tail, while the female is mottled brown for camouflage. Pheasants are omnivores, feeding on seeds, insects, berries, and small invertebrates. They are ground-dwellers but can fly short distances when startled, often bursting into flight with a loud, sudden whirr. During the breeding season, males display and call to attract females, defending their territory with surprising fierceness. As symbols, pheasants remind us that boldness does not always mean loudness, that beauty can be both a lure and a shield, and that sometimes the most powerful presence is the one that chooses exactly when—and how—to be seen.

Pigeon
(Columba livia)

feminine power

The pigeon shares many attributes with its more romanticized cousin, the dove—but its meaning extends further into themes of devotion, navigation, and resilience. In Mesopotamian and Greek mythology, pigeons were sacred to Astarte and Aphrodite, goddesses of love and fertility, linking them to romantic affection, sacred union, and feminine power. In Christian iconography, the dove (a domesticated pigeon) represents the Holy Spirit, divine love, and peace, most famously released by Noah after the flood. In Islamic tradition, pigeons are considered pure and protected, said to have sheltered the Prophet Muhammad during his escape from Mecca. Symbolically, pigeons are also revered for their homing instinct—able to return across vast distances—which makes them enduring symbols of guidance, safe return, and spiritual connection to place. In folklore, pigeons are often seen as messengers between worlds, able to carry news not only across land but between the living and the dead. Spiritually, they represent quiet loyalty, emotional endurance, and the ability to find one's way home no matter how far the journey.

The rock pigeon is native to Europe, North Africa, and western Asia, but has now spread globally due to human settlement. It thrives in urban areas, cliffsides, temples, farms, and city squares, making its home in environments both natural and man-made. Pigeons prefer temperate to warm climates, nesting in crevices, ledges, and rooftops. Known for their soft cooing, iridescent neck feathers, and gentle manner, pigeons are monogamous, often forming lifelong bonds and co-parenting their chicks. They feed on seeds, grains, and scraps, and are intelligent, trainable birds—used throughout history as war-time messengers, spiritual symbols, and even scientific subjects. Their flight is strong, swift, and graceful, and their social nature reflects cooperation, trust, and community resilience. As spirit guides, pigeons remind us to honor our bonds, stay true to our inner compass, and recognize that even the most ordinary presence may carry extraordinary meaning.

Plover
(Charadrius - genus)

maternal devotion

In symbolism and spiritual tradition, the plover is often seen as a guardian spirit, especially in Indigenous American lore, where species like the killdeer are admired for their "broken-wing display"—a behavior used to distract predators from their nests. This remarkable instinct has made the plover a symbol of self-sacrifice, maternal devotion, and the courage to defend the defenseless, even at personal risk. In Celtic and Northern European folklore, the golden plover was seen as a harbinger of changing weather and a messenger between the natural world and the realm of spirits, its flight over the moors echoing the shifting boundaries between life and death. In a broader spiritual sense, the plover represents the ability to survive in exposed places, to find security in vulnerability, and to maintain grace and awareness in wide, open emotional landscapes. It reminds us that you don't have to be big to be brave, and that protection can come through cleverness and calm courage, not just force.

Plovers are found on every continent except Antarctica, favoring coastal habitats, inland wetlands, grasslands, and tundra, depending on the species. They thrive in temperate to tropical climates, often nesting directly on the ground in open areas with little cover, relying on camouflage and behavior to protect their eggs and young. Plovers are insectivorous, using their quick movements and keen vision to snatch prey from mudflats and sandbanks. Many are migratory, such as the Pacific golden plover, which travels extraordinary distances between Alaska and the South Pacific. Their calls are clear and plaintive, often heard at dusk and dawn, echoing across tidal zones and grasslands. As spirit guides, plovers invite us to be vigilant, care deeply, and understand that even in exposed or uncertain circumstances, we can act with wisdom, protect what matters most, and find stillness within openness.

Puffin
(Fratercula - genus)

balance and loyalty

The puffin represents adaptability, balance, loyalty, and transformation. Its life split between land and sea mirrors the balance between emotion and action, home and adventure, inner and outer life. In Celtic and Norse cultures, puffins were seen as coastal protectors and weather spirits, their movements and calls interpreted as signs from the ocean gods. Due to their faithful monogamy and shared parenting, puffins are strong symbols of fidelity, nurturing, and enduring emotional bonds—a quiet reminder that true love is rooted in ritual, presence, and shared effort. In spiritual and modern animal totem interpretations, puffins are seen as emblems of joyful perseverance and soft strength. They teach us that even when the world is harsh and the waves are high, we can dive deep, rise again, and return—carrying nourishment not only for ourselves but for those we love.

Puffins inhabit North Atlantic and Arctic coastlines, from Iceland, Greenland, Norway, and the British Isles to Newfoundland and parts of northern Maine, nesting on rocky cliffs and island outcrops. They thrive in cold marine climates, feeding primarily on small fish like sand eels and herring, which they scoop up in astonishing quantities using their uniquely adapted beaks—holding multiple fish at once without dropping others. The most familiar species is the Atlantic puffin (Fratercula arctica), easily recognized by its bright orange feet, bold black-and-white plumage, and brilliantly colored bill, which fades to gray in the winter and regains its hues in breeding season. Puffins are excellent swimmers and capable flyers, though they appear a bit clumsy in the air—a contrast that only adds to their symbolic theme of duality and gentle contradiction. They nest in burrows or rocky crevices, returning each year to the same mate and nesting site. As spiritual guides, puffins remind us that playfulness and purpose can coexist, that clumsiness can mask great skill, and that we can live lightly on the surface while carrying great depths beneath.

Quail
(Coturnix coturnix)

spiritual guidance

The quail is often associated with safety, nurturing, and the sacredness of home and family. In Native American traditions, particularly among the Pueblo and Hopi, quail are seen as guides of the earth, symbolizing group protection, spiritual guidance in the physical world, and the wisdom of staying low and listening closely. Their tendency to form coveys and move in unison makes them symbols of cooperation, mutual care, and emotional connectivity. In Chinese folklore, quail can represent courage and resilience in battle, often appearing in poetry and paintings as metaphors for quiet bravery—those who act decisively, but without noise or show. Spiritually, the quail is a totem of living simply, moving carefully, and trusting one's inner compass. It reminds us that there is power in small, deliberate steps, and that true strength is often found in those who guard, guide, and nurture from the background.

Quail species are found across North America, Europe, Asia, and parts of Africa, preferring grasslands, shrublands, farms, and dry forests. They thrive in temperate to arid climates, nesting on the ground in well-hidden, shallow scrapes lined with grass. Among the most familiar species are the California quail (Callipepla californica), known for its charming curled head plume, and the Japanese quail (Coturnix japonica), often raised for both eggs and symbolic presence in traditional Asian gardens. Quail are omnivorous, feeding on seeds, insects, berries, and vegetation, and are known for their soft, warbling calls and sudden bursts of flight when startled. They live in tight family groups, with chicks often seen scurrying after parents in neat lines—a visual representation of generational care and unity. As spirit animals, quail teach us to walk with intention, cherish those we travel with, and trust the power of quiet guardianship. They remind us that even the most unassuming among us may be the strongest protectors of what truly matters.

Raven
(Corvus corax)

knowledge

In Norse mythology, the god Odin is accompanied by two ravens, Huginn (thought) and Muninn (memory), who fly across the world and bring him knowledge—making the raven a symbol of divine insight, memory, and the gathering of secret truths. In Native American traditions, especially among Pacific Northwest tribes like the Haida and Tlingit, the raven is a creator, trickster, and transformer, often responsible for bringing light to the world, stealing the sun or moon from a greedy spirit. These stories speak to the raven's role as the bringer of change, both mischievous and miraculous. In Celtic lore, ravens are associated with war and prophecy, especially with goddesses like Morrígan, who appears as a raven on the battlefield. In literature—most famously in Edgar Allan Poe's "The Raven"—the bird becomes a haunting voice of memory, grief, and unresolved longing. Spiritually, the raven is a powerful totem for those who walk between worlds, carrying messages from the subconscious, the spirit realm, or the ancestral past, and teaching us that transformation begins when we learn to honor the dark as deeply as the light.

The common raven is native to the Northern Hemisphere, found throughout North America, Europe, Asia, and parts of North Africa, from tundra and mountains to deserts and forests. They are highly adaptable and extremely intelligent, capable of solving complex problems, using tools, mimicking human speech, and forming strong social bonds. Ravens are often seen alone or in pairs, and they are monogamous, nesting in high cliffs, trees, or urban structures. They feed on everything from carrion and insects to berries and small animals, making them symbolic of survival, resourcefulness, and the ability to thrive in adversity. Their deep, croaking calls and aerial acrobatics give them an eerie, majestic presence. As spiritual guides, ravens invite us to look beneath the surface, trust our inner knowing, and speak truth even when it's uncomfortable. They are guardians of the threshold—between life and death, day and night, known and unknown—and they teach us that true vision comes from seeing in the dark.

Red Kite
(Milvus milvus)

vision and balance

With its deeply forked tail and soaring elegance, the Red Kite (Milvus milvus) is one of Europe's most graceful birds of prey—a creature that drifts effortlessly through the skies with a quiet confidence. Symbolically, the red kite is associated with vision, balance, adaptability, and resurgence. In Welsh and Celtic folklore, it was often seen as a sky guardian, a spirit watching over the land, believed to carry messages between the earthly and ancestral realms. Because it feeds on carrion as well as live prey, the red kite embodies the duality of life and death, loss and renewal, and has become a spiritual emblem of rebirth and healing—especially following its near-extinction and triumphant return in many parts of the UK. Its quiet strength, seen in its wide, circling flights and careful movements, reflects the idea that power does not always need noise—that wisdom often soars above, watching and waiting.

The red kite is native to Western and Central Europe, with strong populations in Spain, Germany, France, and the United Kingdom, where it has been the subject of successful reintroduction efforts. It thrives in temperate climates, preferring a mix of woodland for nesting and open countryside for foraging. This raptor is known for its distinctive appearance: reddish-brown body, pale head, and black-tipped wings, all accented by its iconic tail that twists in flight like a rudder. Red kites feed on carrion, small mammals, birds, and invertebrates, playing an important ecological role as scavengers. Unlike more aggressive birds of prey, red kites are subtle, calculated, and almost silent, often flying in loose, drifting groups over rural hills and meadows. As a totem, the red kite calls on us to trust our instincts, rise above distraction, and find strength in the long arc of recovery. It teaches that sometimes the best way forward is not a charge, but a graceful ascent that sees the full horizon.

Red-Winged Blackbird
(*Agelaius phoeniceus*)

protector of sacred spaces

The Red-Winged Blackbird represents the power of the individual voice, especially in the face of challenge or competition. In Native American traditions, it is sometimes seen as a guardian of the wetlands, a bold spirit that protects sacred spaces and warns of approaching danger. Its colors are rich in meaning: black for mystery and potential, red for passion, vitality, and transformation, and yellow for illumination and clarity. Because males perch conspicuously and sing to assert dominance and attract mates, the bird also speaks to the importance of expression and boundaries —claiming your space, your truth, and your role in the world. Spiritually, it is seen as a herald of seasonal shifts, especially in spring, and as a guide through times of inner growth, creative emergence, or emotional awakening. The red-winged blackbird reminds us that beauty and strength can rise from humble places, and that our voices matter—even when we're one among many in a chorus of life.

Red-winged blackbirds are found throughout North and Central America, thriving in marshes, wetlands, meadows, and agricultural fields, especially in temperate to subtropical climates. Males are boldly colored and vocal, while females are brown and streaked—perfectly camouflaged for nesting in reeds. They feed on insects, seeds, and grains, often foraging in large flocks outside of breeding season. During spring and summer, males fiercely defend their territories, sometimes attacking much larger birds or humans that venture too close to their nests. Their characteristic "conk-la-ree!" song is among the most recognizable sounds of North American marshlands. Social and adaptable, red-winged blackbirds are symbols of community and competition, voice and vigilance. As spirit guides, they encourage you to stand tall, speak clearly, and protect the sacred spaces you hold—within and without.

Rhea
(Rhea pennata)

wisdom and masculinity

A creature of earth wisdom and sacred masculinity, the rhea is deeply tied to fertility, guardianship, and primal instincts. In South American Indigenous mythology, particularly among Patagonian and Andean peoples, the rhea is seen as a nurturer and life-bringer, revered for its unique role in parenting and care—as the males incubate the eggs and raise the young, a rare trait in the bird world. This reversal of roles has made the rhea a symbol of balanced masculinity, where strength includes the ability to protect, nourish, and lead with gentleness. Spiritually, the rhea is a totem of anchored awareness, urging us to stay grounded, defend what we value, and move with quiet purpose rather than spectacle. Its prehistoric appearance also connects it to ancestral memory and survival through vast shifts, making it a guide for those navigating long-term transformation or generational healing.

The rhea is native to South America, with the greater rhea found in Brazil, Argentina, Paraguay, and Bolivia, and the smaller lesser rhea (Rhea pennata) roaming Patagonia and the high Andes. They prefer grasslands, savannahs, shrublands, and open plains, thriving in warm to temperate climates. Rheas are the largest birds in the Western Hemisphere, standing up to 5.6 feet (1.7 m) tall, and though flightless, they are fast runners and powerful kickers when threatened. They are omnivorous, feeding on a mix of plants, fruits, seeds, insects, and small vertebrates. During breeding season, males build and guard nests, incubating eggs from multiple females and fiercely protecting the chicks. Social and nomadic outside of mating periods, rheas may travel in loose flocks. As a symbol, the rhea reminds us that caregiving is an act of power, that vulnerability can coexist with strength, and that rootedness in the earth brings stability in a changing world.

Robin
(Erithacus rubecula)

harbinger of spring

The robin is a harbinger of spring, often appearing just as the earth begins to thaw and life stirs anew—making it a universal sign of renewal, fresh starts, and spiritual awakening. In European folklore, the robin is linked to Christmas and rebirth, often portrayed as a sacred creature that comforted Christ during his crucifixion, earning its red breast from a drop of divine blood. In Celtic and Druidic traditions, robins were seen as spirit messengers, especially from loved ones who had passed on. Their appearance was believed to signify comfort, reassurance, and a reminder that the soul endures. In Native American symbolism, especially in the north, the robin is associated with the sunrise and joy, a symbol of honest work, family harmony, and emotional clarity. As a totem, the robin encourages us to shed the past gently, awaken our inner voice, and find beauty in small acts of kindness and connection.

Robins are found throughout Europe, North America, and parts of Asia, thriving in gardens, forests, fields, and urban parks. They prefer temperate climates, though many migrate to warmer regions in winter. The European robin is small, round, and bold-eyed, with a sweet, clear call and a distinctive orange-red breast, while the American robin, larger and more thrush-like, is known for its cheerful whistling song and its habit of appearing in large flocks during spring migration. Both species are omnivorous, feeding on worms, insects, fruit, and seeds, and are ground-feeders with a curious, approachable nature. They often nest close to human dwellings and are fiercely protective of their young. As spiritual guides, robins remind us that healing can be gentle, that new days always come, and that our voice —however small—can be a song of hope for ourselves and those we love.

Roc
(Middle Eastern Folklore)

fate

The Roc appears most famously in Arabian Nights (One Thousand and One Nights), where the sailor Sinbad encounters the bird during his fantastical voyages. Described as so vast it could block out the sun, the Roc was believed to nest on remote mountains or islands, far beyond the known world. In Persian and Indian mythology, its imagery may stem from earlier sky gods and cosmic birds like the Garuda or Simurgh, making the Roc a symbol of judgment, fate, and unchallengeable natural force. While terrifying in scale, the Roc was not always malevolent—it could also be a protector, a deliverer, or an agent of justice. Spiritually, the Roc represents the power of mythic transformation —those moments in life when we are lifted out of the ordinary and shown the vastness of possibility or peril. It is the archetypal "larger-than-life" force that tests or rescues heroes, reminding us that the universe holds powers far beyond our control, but also that such powers can be navigated with courage and faith.

Of course, the Roc is a mythical creature, not tied to a real-world species, but some scholars believe that its legend may have been inspired by sightings (or exaggerations) of the Aepyornis, the now-extinct elephant bird of Madagascar, which stood over 10 feet tall and laid the largest known bird eggs. Stories told by early travelers and sailors likely magnified these accounts into tales of skies darkened by wings, nests the size of houses, and birds that carried elephants as prey. In this way, the Roc bridges natural wonder and human imagination—its symbolism soaring far beyond zoology into the domain of myth and the subconscious. As a mythic symbol, the Roc reminds us to respect the mysteries of the world, to stay humble in the face of greatness, and to recognize that some truths can only be carried in stories too big to measure.

Roller
(*Coracias garrulus*)

spiritual freedom

The roller is a bird of radiance and rhythm, often associated with creative energy, spiritual freedom, and emotional boldness. Its name comes from its rolling flight patterns—daring aerial flips and dives performed especially during mating displays. In African spiritual symbolism, particularly among Bantu-speaking cultures, the lilac-breasted roller is revered as a sacred sky-dancer, a bringer of good fortune and spiritual vitality. Its bright plumage—shades of blue, green, lilac, and cinnamon—has made it a symbol of beauty through balance, with colors representing different aspects of the soul unified in a single being. In modern spiritual contexts, the roller is seen as a totem of those who live colorfully, speak openly, and embrace risk in pursuit of passion. It encourages self-trust and emotional truth—to roll with life, express joy boldly, and never mute your colors to blend in.

Rollers are found across Africa, Europe, Asia, and parts of the Middle East, thriving in warm, open habitats such as savannahs, scrublands, dry forests, and grasslands. They prefer temperate to tropical climates, often nesting in tree hollows, abandoned woodpecker holes, or earthen banks. The European roller migrates seasonally from Europe to sub-Saharan Africa, while the lilac-breasted roller is a resident bird of eastern and southern Africa, frequently seen perched alone on tree stumps or telephone wires. Rollers are carnivorous, feeding on insects, small reptiles, and amphibians, often caught mid-flight. Their unique flying maneuvers and vividly colored feathers are especially prominent during the breeding season, when males showcase their skills in both flight and brilliance. As symbolic guides, rollers remind us to embrace the dance of life with openness, to speak your truth boldly, and to color your world with authenticity and joy—even if that means standing out against a muted sky.

Rook
(Corvus frugilegus)

prophecy and omen

In European folklore, particularly throughout Celtic, Anglo-Saxon, and Norse traditions, the rook was often regarded as a bird of prophecy and omen, but unlike the solitary raven, the rook was seen more as a communal spirit—a creature connected to family lines, sacred gatherings, and ancestral knowledge. Its habit of nesting in vast colonies called rookeries, often in the same trees year after year, has linked it symbolically to ritual, legacy, and the strength of tradition passed through community. In some old English lore, rooks gathering noisily near a home were said to foretell change, particularly in leadership or spiritual awakening. Yet they are not solely creatures of omen—they're also associated with cleverness, dark humor, and social strategy. Spiritually, the rook invites you to learn from collective memory, to listen to the wisdom echoed through generations, and to recognize that we do not stand alone—we are part of an ancient chorus.

The rook is native to Europe and parts of Asia, thriving in farmlands, open woodlands, meadows, and rural towns, especially in temperate climates. It is distinguished by its bare gray-white patch at the base of its beak, glossy black feathers with a blue-violet sheen, and its deeply social behavior. Rooks feed on insects, seeds, carrion, and scraps, often foraging in large flocks and communicating with a variety of croaks, caws, and clicks. Unlike some other corvids, rooks are almost never seen alone—they are deeply embedded in their groups, working cooperatively and even showing signs of empathy and play. They build elaborate nests high in trees, returning year after year to the same colony. As a symbol, the rook teaches us to honor where we come from, to listen to the voices that came before, and to understand that there is strength not just in intelligence—but in gathered experience, shared purpose, and loyal kinship.

Roseate Spoonbill
(Platalea ajaja)

heart-centered awareness

The Roseate Spoonbill is a creature of emotional cleansing, heart-centered awareness, and spiritual grace. Its delicate pink feathers, like those of the flamingo, are acquired through diet—turning food into beauty—making it a powerful symbol of alchemy, inner change, and becoming radiant from within. The spoon-shaped bill, which it sweeps back and forth to forage in muddy water, speaks to intuition, subtle perception, and sifting through emotion to find sustenance. In spiritual interpretations, the spoonbill represents the ability to turn pain or heaviness into lightness and elegance, to move gracefully through messy, emotional territory, and to emerge with clarity and peace. With its serene posture and soft, dreamlike color, it is a totem of those who heal not by force, but by simply being present, reflective, and open-hearted.

The Roseate Spoonbill is native to the southeastern United States, Central America, the Caribbean, and much of South America, thriving in warm coastal marshes, estuaries, mangroves, and lagoons. It prefers subtropical to tropical climates, where shallow waters offer abundant food. Spoonbills are waders, using their sensitive bills to detect and snap up small crustaceans, fish, and aquatic insects. Their pink coloration comes from carotenoid pigments in their diet, particularly from shrimp and other invertebrates. They often feed and nest in groups, and though they may appear delicate, they are highly adaptable and resilient birds. During flight, their outstretched wings glow with shades of rose, coral, and white, creating a surreal, almost ethereal sight. As a symbol, the roseate spoonbill teaches us to trust our emotional instincts, to find nourishment even in murky waters, and to let beauty be a reflection of what we take in and how we choose to move through the world—with grace, calm, and a soft, steady kind of strength.

Sandpiper
(Calidris minutilla)

messenger of transition

The sandpiper is a messenger of transition, impermanence, and graceful movement through changing emotional landscapes. Because these birds live where earth meets water, they embody the liminal—the spaces where things shift and blur. In Native American traditions, sandpipers are considered spirits of the shoreline, representing alertness, lightness of being, and sacred timing. Their habit of running just ahead of the surf, constantly adjusting, mirrors the idea of moving with life's flow, avoiding overwhelm by dancing just ahead of the tide. In broader spiritual symbolism, they speak to those who are empathetic, observant, and quietly resilient—creatures who sense change before it comes and adapt with humility and grace. The sandpiper teaches us to listen to our surroundings, move gently through uncertain terrain, and recognize that power lies not in force, but in flexibility.

Sandpipers are found all over the world, frequenting coastal shores, estuaries, mudflats, lakesides, and riverbanks, with migratory species traveling thousands of miles each year between breeding and wintering grounds. They thrive in temperate to polar and tropical climates, depending on the species and season. Sandpipers feed on tiny invertebrates like worms, crustaceans, and insects, using their sensitive bills to probe soft earth or sand. Many species nest in the Arctic tundra during summer and migrate south for winter, including the astonishing Sanderling and Red Knot, which undertake epic journeys between continents. These birds are often seen in small flocks, weaving along the shoreline in near-silent coordination. As spirit animals, sandpipers remind us that small movements can lead to great distances, and that sometimes the best way to navigate the world is with quiet confidence, nimble feet, and a deep connection to the rhythm of change.

Sapsucker
(Sphyrapicus - genus)

sacred rhythms

The sapsucker is a bird of sacred rhythms and energetic access. By tapping into the sap of trees, it metaphorically unlocks the life force hidden beneath the surface—making it a totem of those who know how to reach into what seems still and draw out vitality. In Native American traditions, the sapsucker, like other woodpeckers, is seen as a bringer of rhythm, pulse, and life energy, but it holds a special role as a facilitator of nourishment—its holes not only feed itself but also benefit hummingbirds, insects, and other creatures. Spiritually, the sapsucker represents the quiet kind of work that sustains ecosystems, the repeated action that draws sweetness from effort. It reminds us to honor repetition, seek alignment with natural cycles, and recognize that healing or insight often comes from small, consistent efforts rather than dramatic leaps.

Sapsuckers are native to North and Central America, with notable species such as the yellow-bellied sapsucker (Sphyrapicus varius), red-naped sapsucker, and red-breasted sapsucker. They inhabit mixed forests, orchards, and woodlands, particularly in temperate to boreal climates, where they nest in tree cavities and forage for sap and insects. Their feeding technique involves drilling orderly rows of holes into living trees—especially birch, maple, and pine—to access the sugary sap, which they lap up with brush-tipped tongues. These sap wells often become important food sources for other species, making the sapsucker a keystone creature in its habitat. Though not aggressive or showy, they are tenacious, rhythmic, and deeply tuned to the forest's heartbeat. As a spirit guide, the sapsucker calls on us to stay devoted to our process, to tap into deeper energy flows, and to remember that our work—no matter how small—can open pathways for others to thrive.

Scops Owl
(Otus - genus)

otherworldly awareness

The Scops Owl represents subtle guidance, otherworldly awareness, and sacred stillness. In ancient Greek mythology, the Eurasian scops owl (Otus scops) was closely associated with Athena, goddess of wisdom and war strategy. It served as a companion and emblem of her insight—small but all-seeing, silent but deeply knowing. Unlike larger owls that dominate through power, the scops owl symbolizes a more introspective and gentle form of wisdom—the kind that comes from listening to dreams, watching patterns, and understanding the unseen. In European and Mediterranean folklore, the eerie, flute-like call of the scops owl, often heard at dusk or during warm summer nights, was believed to herald messages from the spirit world, or to signal change, initiation, or the need for solitude. As a spiritual guide, the scops owl invites us to embrace the power of the subtle, to trust inner whispers, and to learn from the quiet teachers of the night.

Scops owls are found across Europe, Africa, and Asia, with species such as the Eurasian scops owl, African scops owl (Otus senegalensis), and Indian scops owl (Otus bakkamoena). They live in open woodlands, olive groves, gardens, and savannas, favoring warm, temperate to tropical climates. These owls are small-bodied, typically between 6 and 8 inches (15-20 cm) in height, and are incredibly well-camouflaged with mottled gray, brown, and russet plumage, blending seamlessly into bark or leaves. They feed on insects, small reptiles, and invertebrates, hunting at night with stealth and precision. Their calls are soft, repetitive hoots or whistles—ghostly, rhythmic notes that carry across the dusk. Scops owls are monogamous during the breeding season and often nest in tree hollows. As totems, they remind us to slow down, listen inward, and find guidance in the liminal spaces between light and dark. The scops owl doesn't demand attention—but it sees everything, and speaks only when the message truly matters.

Seagull
(Larus - genus)

the souls of sailors

The seagull represents the seeker, the one who thrives at the edge of things. Because it lives at the meeting point of earth, air, and water, the gull is a creature of liminality, attuned to the ever-changing nature of reality. In Celtic and coastal European folklore, gulls were believed to be the souls of sailors or messengers from the otherworld, returning from distant journeys or carrying news from the beyond. In Native American symbolism, particularly among coastal tribes, seagulls were admired for their sharp vision and fearlessness, often seen as guides and guardians of transitions—warning of storms or signaling change. In modern spiritual interpretations, the seagull has come to represent freedom of thought, clarity, survival instinct, and emotional flexibility. It encourages us to rise above petty worries, see the big picture, and find nourishment even in unlikely places. As a totem, it asks us to consider: are we scavenging for scraps, or are we soaring toward something greater?

Seagulls are found almost everywhere—coastal, inland, and urban environments on every continent—with species adapted to a wide range of temperate to polar climates. While we often associate them with beaches, many gulls also thrive around lakes, rivers, landfills, and city rooftops, showing their remarkable adaptability. They feed on fish, crustaceans, insects, and scavenged food, using problem-solving and keen observational skills. Seagulls are highly social, nesting in large colonies and communicating with an array of vocalizations. They are excellent fliers and opportunistic hunters—sometimes dropping shellfish from heights to break them open or following fishing boats for an easy meal. As spirit guides, seagulls remind us that the ability to shift, glide, and adapt is not a weakness, but a kind of wild wisdom. They teach us that sometimes the edge of the world is exactly where we belong—not to be confined, but to be free, alert, and ready to rise with the wind.

Secretary Bird
(Sagittarius serpentarius)

slayer of evil

The Secretary Bird is a powerful fusion of grace and strength, embodying the idea that true authority comes from clear vision, calm strategy, and decisive action. In African mythology, particularly among Southern and Eastern African tribes, the secretary bird is viewed as a guardian spirit and a slayer of evil, most famously for its ability to hunt and kill venomous snakes. Because it stamps its prey to death rather than using talons like other birds of prey, it symbolizes grounded power, direct action, and intellectual dominance over brute force. In modern symbolism, it has come to represent rational leadership, justice, and the discipline needed to overcome dangerous or toxic influences—whether internal or external. With its quill-like crest feathers said to resemble old scribes' pens, the bird is also associated with wisdom, communication, and the careful handling of knowledge. Spiritually, it teaches us to stay grounded while reaching high, to act swiftly when clarity strikes, and to wield truth like a precise weapon.

The secretary bird is native to sub-Saharan Africa, found in savannahs, grasslands, and open plains, where it prefers wide, unobstructed spaces for walking and hunting. It thrives in warm, dry climates, covering great distances on foot in search of prey, which includes insects, rodents, reptiles, and snakes. Despite its unusual appearance, it is a bird of prey—related to eagles and hawks—and can fly, though it spends most of its time on the ground. Secretary birds are monogamous, nesting in thorny trees where they build large, elevated nests for their chicks. With their towering height (up to 4 feet) and deliberate stride, they command presence in both their natural habitat and cultural imagination. As a symbol, the secretary bird invites us to lead with intellect, protect with calm strength, and stay grounded in both body and purpose, all while maintaining a sharp eye and a fearless heart.

Seriema
(Cariama cristata)

protective spirit

The seriema represents alertness, cunning, and fierce independence. Native to Brazil and neighboring countries, seriemas are deeply woven into folklore and rural wisdom, often admired for their sharp cries that echo across the land, believed by some to warn of danger or signal unseen truths. Because of their ability to hunt snakes and small prey, they are also seen as protective spirits—guardians of the home and countryside, much like the secretary bird of Africa. Their voice, somewhere between a scream and a song, carries a wild, untamed energy, connecting them to themes of freedom, primal instinct, and speaking up when silence is easier. Spiritually, the seriema is a guide for those who must walk alone, move carefully through life's underbrush, and trust their instincts when paths are unclear. It is a totem of those who see what others miss and act when others hesitate.

Seriemas are native to South America, particularly Brazil, Argentina, Paraguay, and Uruguay, thriving in cerrado grasslands, scrubby plains, and rocky savannahs. There are two living species: the more widespread Red-legged Seriema (Cariama cristata), and the more elusive Black-legged Seriema (Chunga burmeisteri). These birds prefer warm, dry climates, spending most of their time on the ground, walking or running through brush in search of insects, lizards, snakes, and small mammals, occasionally supplementing their diet with fruit. Though they can fly short distances, they generally prefer to escape threats on foot, often darting into thick brush. Seriemas are monogamous, nest in trees or on rocky ledges, and are fiercely protective of their young. Their distinctive crest, expressive eyes, and upright stance give them an almost regal, ancient presence, like messengers from another age. As symbolic guides, seriemas teach us to stay watchful, speak our truth, and move forward with clarity and calm—even when the world grows tangled around us.

Shearwater
(Procellariidae - family)

voyager's guide

The shearwater represents emotional resilience, long-haul wisdom, and the ability to navigate through invisible forces with grace. These birds travel vast distances across the open sea, returning only to land for breeding—living their lives largely out of sight, riding the wind above stormy waters. In spiritual terms, the shearwater is a guide for those crossing emotional oceans, a bird that teaches us to trust in unseen rhythms, to conserve energy for what truly matters, and to rely on instinct rather than constant motion. In Polynesian seafaring culture, seabirds like the shearwater were revered as navigational messengers, signaling the presence of islands or guiding voyagers to safe passage. The bird's quiet, fluid flight close to the surface also connects it to the concept of skimming the boundary between conscious and unconscious realms, making it a powerful totem for those who walk between emotional or spiritual depths and outer reality. It's a symbol of patience, clarity in movement, and finding the wind that carries you forward—even when you don't yet see the shore.

Shearwaters are found across all the world's major oceans, with key species like the Cory's Shearwater (Calonectris diomedea), the Manx Shearwater (Puffinus puffinus), and the Short-tailed Shearwater (Ardenna tenuirostris). They thrive in pelagic environments, often flying for thousands of miles between nesting sites and feeding grounds. Shearwaters feed on fish, squid, and crustaceans, often diving or snatching prey from the surface in long gliding flights. Their bodies are built for endurance: long wings, light frames, and a remarkable ability to lock into ocean winds, allowing them to fly across oceans with minimal energy. They return to coastal nesting sites—often on remote islands or cliffs—where they form lifelong pair bonds and raise chicks in burrows. One of the most famous is the Manx Shearwater, which migrates between Wales and South America, covering over 10,000 kilometers each year. As spirit guides, shearwaters remind us that distance does not equal disconnection, that movement can be effortless when aligned with the right current, and that true strength lies not in flapping wildly, but in knowing how and when to glide.

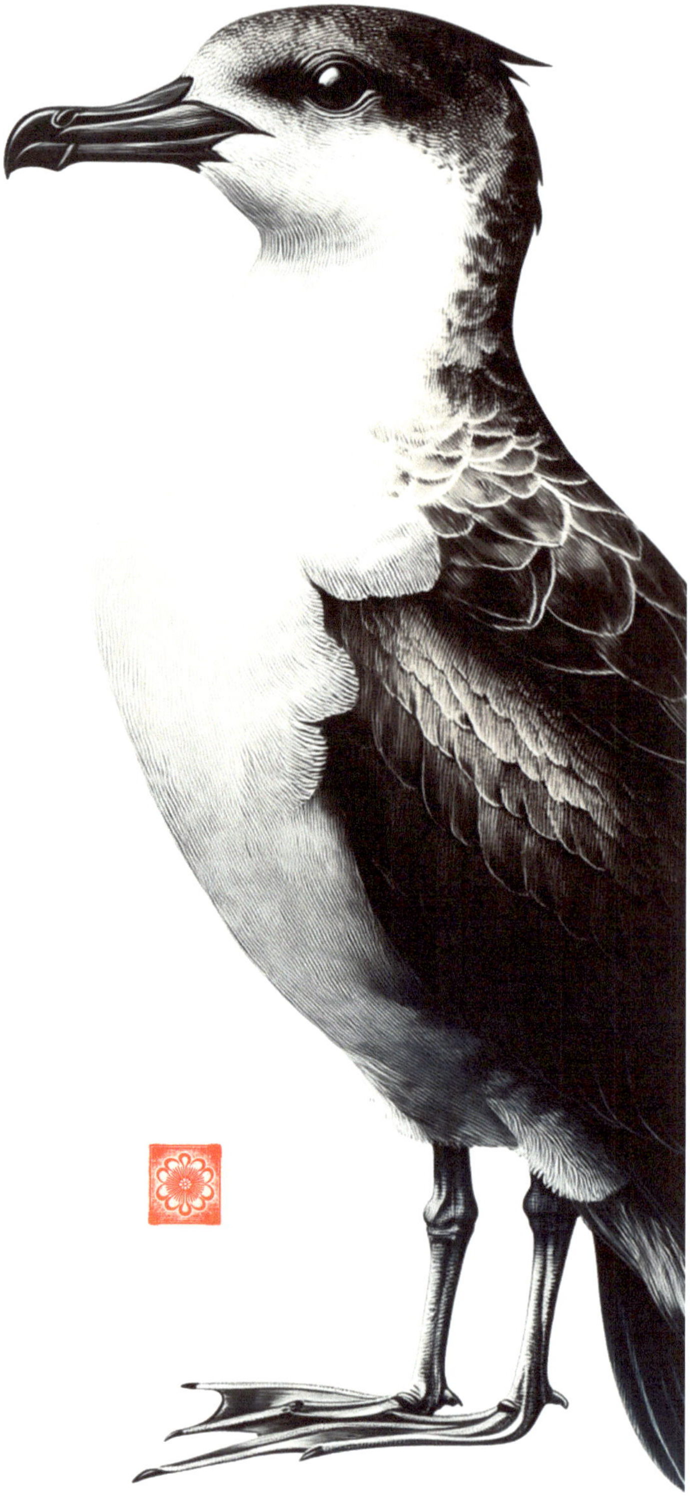

Shrike
(Laniidae - family)

appearances versus reality

The shrike carries intense and layered meanings. Though small and melodious like other passerines, it lives a life more akin to hawks—hunting insects, lizards, mice, and even small birds with swift, efficient precision. This duality makes it a powerful symbol of appearances versus reality, and of strength concealed beneath gentleness. In European folklore, the shrike's habit of leaving its prey pierced on thorns led to associations with justice, sacrifice, and sometimes dark prophecy—as if it were leaving messages in the landscape. In some African and Middle Eastern traditions, shrikes are considered watchers—guardians of boundaries and omens of truth. Spiritually, the shrike reminds us that kindness is not weakness, and that some battles are won with clarity and control, not brute force. It is a totem of sharp perception, assertiveness, and the importance of setting firm boundaries, even when you look harmless.

Shrikes are found across Europe, Asia, Africa, and North America, with notable species like the Loggerhead Shrike (Lanius ludovicianus), Great Grey Shrike (Lanius excubitor), and Red-backed Shrike (Lanius collurio). They thrive in open habitats such as grasslands, farmland, savannahs, and scrubby forests, preferring places with scattered trees or fences where they can perch and scan for prey. Shrikes don't have the strong talons of raptors, so they've evolved the habit of impaling their prey to hold it in place while they feed—behavior that's as strategic as it is grim. They're also known to cache food, creating "larders" on thorn bushes or barbed wire. Shrikes are territorial, solitary, and highly observant, often sitting still for long periods before striking. As spirit animals, shrikes challenge us to own our power, act with purpose, and never underestimate the potency of quiet intensity. They teach that sharpness can be sacred—when it's used to protect, clarify, or cut through the false to reach what's real.

Sparrow
(Passeridae - family)

divine watchfulness

The sparrow represents humility, protection, diligence, and the sacred in the everyday. In Biblical tradition, sparrows are cited as creatures of great value despite their small size—"not one of them falls to the ground outside your Father's care"—making them a symbol of divine watchfulness and love for the overlooked. In Greek and Roman mythology, sparrows were sacred to Aphrodite and Venus, symbols of love, fidelity, and the beauty of ordinary pleasures. In Chinese culture, they have been seen both as symbols of joy and, historically, as pests—yet even that speaks to their deep entwinement with human life, for better or worse. Spiritually, the sparrow is a totem of community and simplicity, urging us to take joy in small things, find strength in cooperation, and remember that no task is too humble when done with care. They teach that gentle persistence and shared effort create lasting foundations, and that in a noisy world, quiet presence is powerful in its own way.

Sparrows are found on every continent except Antarctica, thriving in urban, rural, and wild environments, from temperate to tropical climates. The house sparrow, introduced to many parts of the world, is especially at home around humans, often nesting in the eaves of houses, under roofs, or in garden hedges. Sparrows are omnivores, feeding on seeds, grains, insects, and crumbs, and are known for their chirping vocalizations and social nature, frequently seen in flocks. They're adaptable, prolific, and fiercely protective of their mates and nests. Though often dismissed as common or plain, sparrows embody an everyday kind of nobility—a reminder that the simplest lives are often the most connected, resilient, and meaningful. As spirit guides, sparrows encourage us to nurture our bonds, stay grounded, and take pride in the work that sustains us and those we love.

Spoonbill
(Platalea - genus)

grace and longevity

The spoonbill is a bird of quiet revelation. Its peculiar bill—wide and flat at the tip—is used to sweep through muddy water, feeling for food that it cannot see. This makes the spoonbill a totem of trusting unseen senses, and of drawing nourishment and truth from emotional depths. In spiritual interpretations, the spoonbill encourages us to listen not just with the ears, but with the heart, and to let answers emerge from stillness, not struggle. In Eastern traditions, particularly in Chinese symbolism, birds with long legs and wading habits are often associated with grace, longevity, and dignity, while in more modern ecological symbolism, the spoonbill is seen as a symbol of harmony with water—our emotional and intuitive life—and of balance between movement and reflection. As a spiritual guide, the spoonbill teaches discernment, reminding us that what lies beneath the surface often holds the greatest value.

Spoonbills are found across five continents, with six species distributed in Europe, Asia, Africa, the Americas, and Australasia. They thrive in wetlands, estuaries, mangroves, and shallow lakes, preferring warm temperate to tropical climates. They feed by sweeping their bills side to side, catching small fish, aquatic invertebrates, and plant matter, often foraging in groups with ibis and other waders. Among the most iconic is the Roseate Spoonbill (Platalea ajaja), with its flaming pink plumage and ethereal presence in the Americas, and the Eurasian Spoonbill (Platalea leucorodia), with snowy white feathers and a stately demeanor. Spoonbills are colonial nesters, often building their nests in trees or reeds above water. Though quiet and slow-moving, their persistence and sensitivity allow them to thrive in delicate ecosystems. As symbols, spoonbills remind us that stillness is strength, and that the greatest wisdom often comes when we move with calm intention, stay emotionally attuned, and trust our inner senses to guide us through unclear waters.

Starling
(Sturnus vulgaris)

messenger bird

The starling represents unity through diversity, the magic of group coordination, and the power of shared purpose. In Celtic and European folklore, starlings were considered messenger birds, especially associated with omens, prophecy, and social cohesion. Their uncanny ability to move in massive, perfectly synchronized flocks —turning, twisting, and flowing through the sky like a single organism—makes them spiritual emblems of connectedness, intuition, and the flow of information between individuals and the larger collective. In modern symbolic interpretations, the starling teaches us about cooperation without loss of individuality, the importance of listening and responding rather than dominating, and the idea that beauty is born when we move together, not against one another. It is also a bird of vocal mimicry, often replicating the sounds of other birds or even mechanical noises—making it a totem of expression, language, and adaptation.

Starlings are native to Europe, Asia, and North Africa, though the common starling (Sturnus vulgaris) has been introduced widely, including across North America, Australia, and New Zealand. They thrive in a variety of temperate climates, especially in farmlands, urban areas, open woods, and grasslands, where they can forage for insects, fruits, seeds, and grains. Their feathers are often iridescent, shimmering with purples and greens in the right light, despite their dark overall appearance. Highly social and intelligent, starlings nest in cavities and roost in enormous communal flocks—sometimes numbering in the millions. Their murmuration flight patterns are still studied today as models of complex systems and decentralized coordination. As spirit animals, starlings remind us to value the role we play within a larger whole, to find harmony in movement, and to honor communication not just as noise, but as the lifeblood of connection and change.

Stilt
(Himantopus - genus)

emotional equilibrium

The stilt is often seen as a bridge between realms—a creature that walks the edge of water and land, grounded but deeply attuned to fluidity. Its long legs and slender profile give it an appearance of near-weightlessness, making it a symbol of emotional equilibrium, spiritual poise, and heightened awareness. In some Indigenous traditions and folklore from Oceania and Asia, wading birds like the stilt are viewed as watchers of the in-between—those who navigate dreams, tides, and transitions, offering messages during periods of change or emotional reflection. Spiritually, the stilt encourages us to move carefully through emotional landscapes, to stay centered even when things feel unstable, and to recognize that there is power in stillness, alertness, and quiet discernment. It is a totem of those who walk delicate paths—who must balance their sensitivity with confidence and remain poised while emotions swirl around them.

Stilts are found across warm and temperate regions on every continent except Antarctica, with notable species including the Black-winged Stilt (Himantopus himantopus), Black-necked Stilt (Himantopus mexicanus), and Pied Stilt (Himantopus leucocephalus). They thrive in wetlands, estuaries, salt flats, shallow lakes, and flooded fields, where their long legs allow them to wade through water too deep for most birds their size. Stilts feed on insects, crustaceans, mollusks, and small aquatic creatures, using their long, fine bills to pick delicately through the surface or just below it. They are often seen in pairs or loose flocks and are protective and vocal when guarding nests, which are typically shallow scrapes on the ground. Despite their fragile appearance, stilts are determined, loyal, and surprisingly bold when defending their territory. As spirit guides, stilts remind us to move with elegance even in emotionally murky waters, to honor our sensitivity, and to remain balanced and aware, no matter how uncertain the terrain may seem.

Stork
(Ciconiidae - family)

new life

The stork is most famously associated with new life and safe delivery, a reputation rooted in European folklore where storks were believed to deliver babies to homes, nesting atop chimneys as a sign of fertility and protection. This association likely emerged from the stork's strong monogamous pair bonds and its habit of returning to the same nest year after year—making it a symbol of domestic stability, familial love, and life's return after dormancy. In Egyptian mythology, the stork was linked to the Ba, or the soul, thought to leave the body during sleep and return upon waking—connecting the bird to the journey of the spirit and the cycles of consciousness. In Chinese symbolism, the stork represents longevity, good fortune, and filial piety, often depicted flying heavenward with a soul or descending to bless a family. Spiritually, the stork serves as a guide through life's thresholds—birth, death, rebirth—and teaches us to honor cycles, remain faithful to our path, and move with grace between worlds.

Storks are found throughout Europe, Asia, Africa, and parts of the Americas, with species such as the White Stork (Ciconia ciconia), Black Stork (Ciconia nigra), and the enormous Marabou Stork (Leptoptilos crumenifer) of Africa. They inhabit wetlands, river valleys, grasslands, and agricultural areas, preferring warm to temperate climates. Storks are carnivorous, feeding on fish, frogs, insects, reptiles, and small mammals, and are often seen walking methodically through shallow water or tall grass in search of prey. They are migratory, with some species flying thousands of miles between breeding and wintering grounds, soaring on thermal currents with minimal effort. Storks build large nests, often reused and added to over many years, symbolizing endurance and the sacred hearth. As spirit animals, storks encourage us to honor our roots, trust in life's timing, and carry our purpose with quiet dignity. They teach us that even in long migrations, we are always flying toward home.

Swallow
(Hirundinidae - family)

harbinger of spring

The swallow carries a timeless association with homecoming and protection, often seen as a harbinger of spring and renewal. In ancient Greek and Roman culture, swallows were sacred to goddesses like Aphrodite and Venus, representing love, transformation, and resurrection. In Christian tradition, they symbolized rebirth and the triumph of the soul, frequently appearing in religious art as messengers of divine grace. In sailor folklore, a tattoo of a swallow signified that a seafarer had traveled over 5,000 nautical miles—a symbol of safe return, loyalty to one's crew, and a spiritual guide who would carry the soul home if lost at sea. Swallows build their nests in the eaves of homes and barns, often returning to the same spot year after year, reinforcing their role as a guardian of family, memory, and continuity. As a spirit animal, the swallow represents hope renewed, faith after hardship, and the belief that beauty and connection will always find their way back to us.

Swallows are found across every continent except Antarctica, thriving in open landscapes near water, farmlands, and human dwellings, particularly in temperate to tropical climates. They are insectivores, catching flies, mosquitoes, and other small insects mid-flight with breathtaking agility. Among the most familiar species are the barn swallow (Hirundo rustica), with its glistening blue back and long forked tail, and the cliff swallow, known for its communal nesting habits under bridges and rock faces. Swallows are migratory marvels, traveling thousands of miles between breeding and wintering grounds, often navigating with uncanny accuracy to the very site where they were born. They are deeply social, often seen in flocks during migration or resting in dense, chattering groups. As totems, swallows remind us to trust in the cyclical nature of life, to carry love and light wherever we go, and to always return to what truly matters—no matter how far we fly.

Swan
(Hirundinidae - family)

divine harmony

The swan holds a sacred place in myth, poetry, and spiritual tradition. In Greek mythology, it was sacred to Apollo and Aphrodite, representing poetic inspiration, divine harmony, and beauty born of both joy and sorrow. The swan's association with music and art is echoed in the ancient belief in the "swan song"—a final burst of beauty or truth before death, turning the bird into a symbol of epiphany, spiritual awakening, and letting go with grace. In Celtic lore, swans were messengers from the Otherworld, often taking on human forms or guiding souls across watery thresholds, tying them to transformation, reincarnation, and ancestral presence. In Hinduism, the swan (or hamsa) is a symbol of spiritual discernment and liberation, said to be able to separate milk from water—representing the ability to perceive truth amid illusion. As spirit animals, swans embody love in its most soul-deep form: faithful, serene, and enduring. They call us to embrace our elegance, move through emotional waters without fear, and recognize the divine within the quiet moments of life.

Swans are found across Europe, Asia, North America, and Australia, thriving in lakes, rivers, wetlands, and coastal bays. They prefer temperate climates and are especially associated with serene, reflective environments. The mute swan (Cygnus olor), iconic in Europe with its orange bill and poised silence, is among the most recognized species, while the trumpeter swan and whooper swan are celebrated for their haunting, resonant calls. Swans are largely herbivorous, feeding on aquatic plants, and are known for their monogamous pair bonds, often mating for life and raising their cygnets with attentive devotion. Their elegant flight and fierce protectiveness make them symbols of both grace and strength, of beauty that guards, defends, and endures. As spiritual guides, swans encourage us to trust the currents of life, to hold space for deep love and sacred silence, and to move through emotional transformation with the dignity of a creature who belongs equally to this world and something far beyond it.

Swift
(Apus apus)

spiritual ascension

The swift represents spiritual ascension, focus, and the pursuit of higher truth. In cultures across Europe, Asia, and Africa, swifts have long been associated with the sky spirits, souls of the dead, and messengers from the divine, due to their almost constant presence in the air and their rarely seen landing behavior. The swift's ability to sleep, eat, and mate while flying makes it a creature of liminality and ethereal existence—it lives between worlds, touching the earth only to breed and then vanishing into skybound silence. In many modern spiritual interpretations, swifts symbolize the importance of living in alignment with one's purpose, of not wasting energy on what grounds you unnecessarily, and of embracing movement, clarity, and freedom. They are also deeply tied to timing and seasons—their return often signals the arrival of summer, making them omens of change, warmth, and creative flow. As spirit guides, swifts remind us to rise above distractions, trust our instincts, and know that what is fleeting may still carry eternal meaning.

Swifts are found across Europe, Asia, Africa, and the Americas, thriving in open skies above forests, cities, cliffs, and wetlands, especially in temperate to tropical climates. The Common Swift (Apus apus) is famous for spending up to 10 months in uninterrupted flight, gliding and darting with astounding precision. They feed entirely on aerial insects, forming large, high-flying groups during migration and feeding. Unlike swallows, they do not perch but cling vertically to walls, trees, or cliffs when nesting. Swifts often return year after year to the same nesting sites, forming lifelong pair bonds and showing remarkable navigational memory. Their flight is exhilarating and acrobatic, cutting through the air with sharp calls and impossible turns. As a symbol, the swift calls us to embrace the rhythm of movement, to let go of the need to settle too soon, and to live fully in motion—driven by instinct, spirit, and the pursuit of something beyond the visible horizon.

Tanager
(Thraupidae - family)

blessings of the forest

The tanager is associated with colorful self-expression, emotional release, and spiritual uplift. In South and Central American folklore, tanagers—especially the scarlet tanager and painted bunting—are seen as blessings of the forest, messengers of joy and celebration. Their dazzling feathers and sweet, flutelike songs link them to the heart chakra, encouraging those who encounter them to open themselves to love, artistic inspiration, and emotional authenticity. Because they often appear suddenly, like living jewels darting through the trees, they also symbolize surprise blessings and unexpected joy—reminders that beauty and peace can show up even in the midst of chaos or routine. As spiritual totems, tanagers invite us to reconnect with wonder, to embrace color in both thought and feeling, and to let our true nature shine unapologetically, without fear of being seen.

Tanagers are native to the Americas, particularly abundant in the tropical forests of Central and South America, though some species—like the scarlet tanager (Piranga olivacea)—migrate into North America during the summer months. They inhabit woodlands, rainforests, gardens, and shrubby edges, favoring warm, humid climates, though some adapt well to higher elevations and seasonal migrations. Their diet includes fruits, berries, nectar, and insects, which often influences the intensity of their coloration—making them symbols of how what we consume affects how we shine. Males are typically more brightly colored, especially during breeding season, while females are subtler, blending more into the foliage. Despite their delicate appearance, tanagers are alert and adaptable, thriving in the layered complexity of forest life. As spirit guides, they encourage you to sing your own song, let your emotions be expressed through beauty and action, and to find healing not in escape, but in embracing the full spectrum of your inner world.

Tern
(Laridae - family)

messenger of the wind

The tern represents guidance, fidelity, and unwavering intention. It is often seen as a messenger of the wind, navigating vast oceans with lightness and focus, embodying the idea of holding your course even when the destination is far beyond the horizon. In many coastal and maritime traditions, the tern is regarded as a sign of safe passage and soul travel, admired for its precision in flight and loyalty to its breeding grounds. Because some species—like the Arctic Tern—migrate from the Arctic to the Antarctic and back again each year (the longest migration of any animal on Earth), the tern also symbolizes endurance, spiritual pilgrimage, and the cycle of returning to one's true center no matter the distance traveled. Spiritually, the tern encourages us to trust the winds of life, to fly lightly with faith, and to remain focused on what matters most, no matter how far or unreachable it may seem.

Terns are found on every continent, especially near coastal waters, estuaries, islands, and inland lakes, thriving in temperate to polar climates, depending on the species. They feed primarily on small fish and marine invertebrates, diving headfirst into the water with astonishing speed and accuracy. Their slim bodies and strong wings make them exceptional flyers, often hovering above water before striking. The Arctic Tern (Sterna paradisaea), for instance, makes an annual round-trip of over 44,000 miles, chasing summer light from pole to pole—making it not only a symbol of freedom, but of eternal return and purposeful flight. Terns are monogamous, often forming long-term bonds and returning to the same nesting sites each year. Their protective nature toward their nests and young also connects them to loyalty, caregiving, and sacred guardianship. As totems, terns guide us to seek clarity over noise, to move through emotional or spiritual landscapes with grace, and to stay devoted to our path, even when the waters beneath us are deep and unknown.

Thrasher
(Toxostoma - genus)

voice of the inner world

The thrasher represents the release of buried emotion and the importance of giving voice to the inner world. Named for its habit of "thrashing" through leaves and debris to uncover hidden food, the bird is a totem of emotional digging and spiritual unearthing—encouraging you to sift through the clutter of your past or your environment to find what still nourishes your spirit. In Native American lore, particularly in regions of the American Southwest, the thrasher's persistent foraging and robust song are seen as traits of determination, inner work, and authenticity. The Brown Thrasher (Toxostoma rufum), known for its astonishing vocal repertoire—over 1,000 distinct song variations—has become a symbol of creative intelligence, verbal dexterity, and personal storytelling. Spiritually, the thrasher urges us to speak our truth clearly, to dig beneath the surface of situations or emotions, and to turn messiness into meaning.

Thrashers are native to North and Central America, especially found in scrublands, woodlands, deserts, and overgrown fields, preferring warm to temperate climates. They are medium-sized songbirds with long tails, downward-curving bills, and often streaked plumage that blends into dense vegetation. Their diet includes insects, fruits, seeds, and small invertebrates, which they search for by flipping leaves and digging with precise, repeated motions. Despite their secretive nature, thrashers are not shy when it comes to song—singing from exposed perches in spring and summer with a loud, richly varied series of whistles, warbles, and phrases. The Brown Thrasher, in particular, is often solitary and territorial, yet its voice carries through the underbrush like a poet reciting to the wild. As a symbolic guide, the thrasher reminds us that creativity is born from chaos, that expression requires depth, and that sometimes we must rustle the leaves of the past to sing clearly into the future.

Thrush
(Turdidae - family)

otherworldly wisdom

The thrush is a bird of inner truth, poetic expression, and connection to the divine through sound and solitude. In Celtic mythology, the thrush was said to sing with otherworldly wisdom, its song guiding listeners into enchanted realms or deeper understanding. The hermit thrush, native to North America, has long been considered a spiritual messenger, its pure, spiraling notes representing the voice of the forest or the whisper of the soul. Poets like Walt Whitman and Seamus Heaney referenced the thrush in their work as a symbol of nature's grace in grief and the healing power of quiet music. In modern symbolism, the thrush is a totem of gentle transformation—a reminder to listen carefully, reflect deeply, and speak when the heart is ready. It calls us to find harmony between solitude and song, and to embrace the stillness from which inspiration arises.

Thrushes are found across Europe, North and South America, Asia, and parts of Africa, thriving in woodlands, parks, gardens, and forest edges, particularly in temperate and subtropical climates. These birds are typically medium-sized, with streaked or spotted breasts, muted earth-toned plumage, and upright, attentive posture. They are omnivorous, feeding on insects, worms, berries, and fruits, often foraging by hopping quietly along the ground. The wood thrush (Hylocichla mustelina) and song thrush (Turdus philomelos) are especially known for their rich, layered songs, which they repeat with variation and phrasing that resembles a musical composition. Many species are migratory, and their return often coincides with spring's awakening. As spiritual guides, thrushes remind us that great power can lie in soft voices, that truth doesn't need to shout, and that there is magic in simplicity, rhythm, and natural expression.

Titmouse
(Turdidae - family)

messengers of happiness

The titmouse is a creature of curiosity, simplicity, and emotional lightness. Though small, it is fearless—often the first bird to investigate something new in its environment or to flit near humans. In Native American traditions, especially among woodland cultures, small birds like the titmouse are seen as messengers of happiness and play, bringing news of good fortune and encouraging people to stay alert to the magic in the mundane. Spiritually, the titmouse represents the ability to remain joyful and inquisitive even in the face of challenge, teaching that small gestures of kindness and wonder carry enormous strength. It reminds us to be light without being shallow, and to move through life with trust in the rhythms of community, laughter, and gentle connection. As a totem, it encourages confidence without arrogance, presence without pressure, and the belief that joy can be both a practice and a power.

The titmouse is found primarily in North and Central America, especially across the eastern and southeastern United States, thriving in deciduous forests, woodland edges, gardens, and suburban areas. The Tufted Titmouse is easy to recognize by its soft gray feathers, white belly, and distinct crest, often paired with a sweet, whistled "peter-peter-peter" song. They feed on insects, seeds, berries, and nuts, often seen hanging upside-down from branches or joining mixed-species flocks in winter. Titmice are monogamous, nesting in cavities or birdhouses, and are known for caching food and showing clever foraging behaviors. Despite their small size, they are assertive at feeders and alert in the wild, never afraid to claim their space. As symbolic guides, titmice teach us to live with curiosity, speak kindly, and never underestimate the strength found in joy, community, and an open, watchful heart.

Toucan
(Ramphastos toco)

physical and spiritual realms

The toucan is a bird of bold voice and brilliant visibility. Its massive bill, while seemingly cumbersome, is actually lightweight and highly useful—an emblem of how something that stands out can also serve a vital purpose. In Amazonian mythology, the toucan is considered a sacred bird, often associated with the spirit world, tribal shamans, and the rainbow gods. Its striking appearance and echoing call make it a bridge between the physical and spiritual realms, a messenger who draws attention to truths that need to be spoken. Spiritually, the toucan represents confidence in one's voice, pride in one's identity, and the courage to live vibrantly and truthfully. It calls us to embrace what makes us unique, to speak our truths without fear, and to use our visibility as a way to inspire and uplift others.

Toucans are native to tropical and subtropical rainforests of Central and South America, where they dwell in the canopy layer, often seen hopping along branches or flying short distances between trees. They thrive in warm, humid climates, feeding primarily on fruit, though they also eat insects, small reptiles, and bird eggs. Their massive bills, often longer than their heads, are used not only for feeding but also for regulating body temperature and engaging in playful social behavior. Among the most famous species is the Toco Toucan, the largest of its kind, with its striking orange bill and white throat patch. Toucans are social, intelligent, and vocal, often seen in small groups, communicating with croaking, clattering calls that echo through the rainforest. As spiritual guides, toucans remind us to celebrate color and sound, embrace our individuality without shame, and communicate with boldness, authenticity, and joy—even in a world that often rewards silence and conformity.

Turkey
(Meleagris gallopavo)

giver of life

The turkey is a powerful emblem of sacred generosity and Earth-centered spirituality. In many Native American traditions, particularly among tribes like the Hopi, Cherokee, and Pueblo peoples, the turkey is revered as a giver of life, honored for its willingness to provide for the tribe and for its connection to agricultural cycles and seasonal rhythm. It embodies the idea that true abundance comes from sharing, not hoarding, and that to live in balance with the earth is to live in respect for the cycles of giving and receiving. Spiritually, the turkey is a totem of gratitude, resourcefulness, and the grounding power of community, reminding us to honor our roots, be generous with our gifts, and find fulfillment not in excess, but in harmony. While not flashy or predatory, the turkey's power lies in its presence, resilience, and its role as a symbol of sustenance and spiritual humility.

Wild turkeys are native to North America, thriving in woodlands, meadows, and grasslands across the United States, Mexico, and parts of Canada. They prefer temperate climates and are incredibly adaptable—capable of living in forests, scrub, and even near farms. Despite their bulky appearance, wild turkeys are strong, agile birds that can run at high speeds and fly short bursts when startled. They forage on the ground for acorns, seeds, berries, insects, and small reptiles, and are highly social, often moving in flocks. Turkeys are also vocal and intelligent, with a variety of calls, postures, and displays, especially during mating season. The male's fanned tail and gobble are iconic symbols of masculine display and seasonal celebration, especially around autumn. As spiritual guides, turkeys encourage us to walk humbly, live generously, and remember that true wealth is found in relationships, tradition, and the land beneath our feet.

Vulture
(Accipitridae - family)

spiritual recycling

The vulture is a creature of thresholds—neither predator nor prey, but a cleanser of death, turning decay into continuity. In Ancient Egyptian mythology, the vulture was a symbol of divine motherhood and protection, often associated with the goddess Nekhbet, a protector of pharaohs and sacred transitions. In Tibetan and Hindu traditions, vultures play a literal role in sky burials, consuming human remains to help return the body to nature, making them sacred agents of release, detachment, and spiritual recycling. In Native American lore, vultures are respected as purifiers, never killing but always cleaning, associated with the element of air and the unseen realm, reminding us that what seems ugly may serve a higher purpose. Spiritually, the vulture teaches patience, clarity, and the power of waiting for the right moment. It asks us to honor what must end, to see beauty in the breakdown, and to understand that true transformation often begins in stillness and surrender.

Vultures are found on every continent except Australia and Antarctica, thriving in open plains, deserts, savannahs, and mountainous regions. There are two major types: Old World vultures (like the griffon vulture) in Europe, Asia, and Africa, and New World vultures (like the turkey vulture and Andean condor) in the Americas. They thrive in warm, arid to temperate climates, using their exceptional eyesight and smell (in the case of New World species) to find carrion. Vultures are highly social and often feed communally, circling the sky in silent grace. Their bald heads are evolutionary adaptations for cleanliness, and despite their grim reputation, vultures do not harm the living—they simply fulfill a vital ecological role. As spirit animals, vultures ask us to redefine beauty, let go of judgment, and recognize the sacred in what others reject. They remind us that death is not the end, but a necessary part of the eternal cycle.

Wagtail
(Motacilla - genus)

cosmic rhythm

The wagtail is a creature of lightness and balance, always in motion but never in chaos. In Japanese folklore, wagtails (sekirei) are associated with creation myths, believed to have taught the first gods how to perform the sacred mating ritual, thus linking the bird to life, harmony, and cosmic rhythm. In Celtic tradition, they were seen as omens of change and progress, their constant motion representing the soul's journey forward —even in small steps. The wagtail's signature tail-bobbing has made it a symbol of playfulness, spiritual rhythm, and attentiveness to signs in the natural world. Spiritually, the wagtail is a totem of those who may seem delicate but are incredibly attuned, reminding us to move lightly, follow the water's edge, and trust in the power of small, consistent actions to bring lasting joy and transformation.

Wagtails are found across Europe, Asia, and Africa, with some species extending into parts of Australia and Alaska, thriving in open countryside, near water, meadows, riverbanks, and farmland. They prefer temperate to subtropical climates, often nesting on the ground or in low walls and structures. The Pied Wagtail (Motacilla alba), with its black-and-white plumage and brisk movements, is especially beloved in the British Isles, while the Grey Wagtail (Motacilla cinerea) and Yellow Wagtail (Motacilla flava) bring flashes of color to streams and meadows across Europe and Asia. These birds are insectivores, often seen darting after flies and gnats in midair, always alert, agile, and perfectly balanced. As spirit guides, wagtails encourage us to stay in motion—not frantic, but flowing—to delight in life's small dances, and to remember that even the lightest touch, when given rhythm and intention, can ripple through the world like water.

Warbler
(Sylviidae- family)

divine whispers

The warbler is a bird of hidden truth and poetic presence. It is rarely bold or dramatic, but it sings intricate, soulful songs, often from deep within dense foliage, reminding us that not all wisdom seeks attention. In folk traditions across Europe and North America, warblers were believed to foretell the coming of spring, rain, or emotional renewal, their songs viewed as divine whispers or omens of joy in moments of stillness. Because warblers are difficult to spot but often heard, they are symbolic of intuition and subtle influence, teaching that what is unseen can still be deeply felt. Spiritually, the warbler is a totem of those who communicate through feeling, art, and quiet insight, showing that a soft voice can resonate deeply, and that small gestures can hold profound meaning.

Warblers are found across Europe, Asia, Africa, and the Americas, thriving in forests, shrublands, wetlands, gardens, and along riverbanks, especially in temperate and tropical climates. Species like the wood warbler (Phylloscopus sibilatrix), yellow warbler (Setophaga petechia), and blackcap (Sylvia atricapilla) are beloved for their melodic, complex calls, often delivered from hidden perches. Most warblers are migratory, undertaking long journeys between breeding and wintering grounds, and many are insectivorous, feeding on midges, caterpillars, and small arthropods. Their plumage is often subtle—olive, yellow, gray—but their songs are anything but plain. Despite their size, warblers are tireless travelers and fiercely focused during their short nesting season. As spirit guides, they remind us to embrace our sensitivity, to trust the unseen rhythms of life, and to know that sometimes the most beautiful truths are whispered, not shouted.

Waxwing
(Bombycilla - genus)

unity and sharing

The waxwing is a creature of grace, ephemerality, and community spirit. Unlike many songbirds, waxwings don't defend territories or sing complex solos—instead, they travel and feed together in cooperative, peaceful flocks, making them powerful emblems of unity, sharing, and non-competitive coexistence. In Indigenous North American traditions, particularly among northern tribes who observe the Cedar Waxwing's winter flocks, the bird is seen as a harbinger of abundance and joy, especially in lean seasons. Spiritually, the waxwing is often considered a messenger of present-moment pleasure, a reminder to savor what is fleeting—beauty, affection, sweetness—without clinging or hoarding. Their preference for fermented berries (sometimes resulting in slightly tipsy behavior) also gives them a lighthearted, almost Bacchic symbolism, linking them to celebration, intoxication, and the idea that joy shared is joy multiplied. As totems, waxwings encourage emotional generosity, mindful enjoyment, and the grace to move softly through the world in tune with others.

Waxwings are native to North America, Europe, and northern Asia, thriving in boreal forests, woodlands, orchards, and gardens, especially in cool to cold temperate climates. They are nomadic, often traveling in flocks in search of fruiting trees and shrubs. The Cedar Waxwing (Bombycilla cedrorum), common in North America, is particularly known for its sleek, pale brown-and-gray plumage, lemon-yellow tail tip, and subtle red wax-like wing tips, whose function remains a mystery but adds to their ethereal appearance. They feed primarily on berries and small fruits, occasionally supplementing their diet with insects. Because they often feed each other berries during courtship and flock behavior, they symbolize gentle affection, communal care, and the spiritual act of giving without expectation. As spiritual guides, waxwings invite us to cherish beauty without grasping it, to walk kindly beside others, and to celebrate life's sweetness while it lasts.

Weaver
(Ploceus cucullatus)

interconnectedness

The weaver represents the interconnectedness of effort and vision, the belief that what we create—whether a home, a relationship, or a purpose—matters when made with care and clarity. In African and South Asian traditions, weavers are admired not only for their beauty but for their artistry and persistence, constructing complex nests that sway like lanterns from tree branches or reeds. Their structures often serve as communal signals—symbols of stability, fertility, and order. In spiritual terms, the weaver is a totem of designing one's destiny, reminding us that our lives are woven thread by thread from intention, environment, and action. The male weaver's building of a nest to attract a mate also symbolizes offering, effort, and the value of showing care through what we build. They call us to ask: What am I creating with my energy? Does it shelter love, invite beauty, or reflect my truth?

Weavers are native to sub-Saharan Africa, South Asia, and parts of Southeast Asia, often found in savannahs, wetlands, grasslands, and agricultural areas, where they thrive in warm, dry to humid tropical climates. Most weavers are seed-eaters, often foraging in flocks, and their striking yellow, black, or russet plumage—especially among males during breeding season—makes them stand out like flying sparks in the reeds. Their nests are marvels of avian engineering: woven spheres or ovals with side entrances, made from grasses, palm strips, or even human materials like string and plastic. Males often build multiple nests to attract females, whose inspection determines mating. Despite their size, weavers are fierce defenders of their colonies, often nesting close together in vast, chattering groups. As spirit animals, weavers teach us to honor the work of our hands, to find power in precision and persistence, and to remember that creativity is not just for expression—but for shelter, love, and legacy.

Whip-poor-will
(Antrostomus vociferus)

guardian of thresholds

The whip-poor-will is a guardian of thresholds—those spaces between day and night, waking and dreaming, life and death. In Appalachian and Native American folklore, the whip-poor-will is both revered and feared, thought to steal a person's soul when it sings near a house or to foretell death, much like the banshee in Celtic myth. Yet its voice, ethereal and rhythmic, is also considered a song of the spirit world, connecting listeners to ancestors, lost loved ones, and deeper emotional truths. Its name, echoing into the dusk like a spell, is a reminder that what is repetitive may not be mindless—it may be sacred ritual or remembrance. Spiritually, the whip-poor-will is a totem of dreamers, poets, mystics, and wanderers—those who navigate life's shadows and learn to listen for messages in silence, solitude, and sound.

Whip-poor-wills are native to eastern and central North America, favoring deciduous forests, pine woods, and open glades. They thrive in temperate climates, especially near forest edges, where they can forage and sing from the cover of leaves. These birds are insectivores, feeding on moths, beetles, and flying insects, which they catch mid-flight with their wide, gaping mouths during the twilight and night hours. Their camouflaged plumage blends seamlessly into leaf litter and bark, making them almost invisible by day. Unlike most songbirds, they nest directly on the forest floor, laying eggs among dry leaves without building a structure. Their song, endlessly repeating their name, may continue for hundreds of calls in a row, creating an atmosphere of trance or invocation. As spirit guides, whip-poor-wills remind us to trust what can't be seen, to honor the spaces between endings and beginnings, and to find meaning in the quiet music of the dusk—where memory, mystery, and moonlight intertwine.

Woodpecker
(Dryocopus pileatus)

ancestral protection

The woodpecker is a master of persistence and inner calling. Its distinctive drumming—used for communication, territory marking, and foraging—makes it a powerful symbol of focus, timing, and the importance of listening to your own inner rhythm. In Native American traditions, the woodpecker is often considered a guardian of the heartbeat, associated with earth energy, ancestral protection, and the sound of creation itself. Some tribes believed its rhythmic pecking was a way of awakening spiritual truths or hidden knowledge within trees, the symbolic keepers of wisdom. In European folklore, woodpeckers were messengers between worlds, sometimes omens of approaching change or protectors of sacred forests. Spiritually, the woodpecker is a totem for those who dig deep, who trust the beat of their intuition, and who carve out their own space with determination and precision. It reminds us that opportunity may be hidden beneath the surface—but those who persist will unlock it.

Woodpeckers are found across North and South America, Europe, Africa, and Asia, living in forests, woodlands, parks, and even deserts, depending on the species. They thrive in temperate to tropical climates, feeding primarily on insects, larvae, and tree-dwelling invertebrates, as well as fruits and nuts. Their specially adapted skulls and shock-absorbent beaks allow them to peck into trees without harm—symbolizing the ability to break through hard barriers without damage to the self. They often excavate nesting cavities in dead or decaying trees, providing shelter for many other species, making them symbols of ecosystem builders and silent benefactors. Some, like the acorn woodpecker, even create food caches—complex storage systems that speak to their intelligence and foresight. As spiritual guides, woodpeckers encourage us to act with precision, follow our instincts, and remember that steady, purposeful action creates the rhythm of real change.

Wren
(Troglodytidae - family)

sacred mystery

The wren is a bird of quiet strength, cleverness, and sacred mystery. In Celtic and Druidic mythology, the wren was called the "king of all birds", not for its size, but for its wisdom and cunning—a title earned in myth when it hid on the back of an eagle and soared higher than any bird, using strategy over strength. In Irish tradition, it is deeply associated with song, prophecy, and the turning of the year, famously linked to St. Stephen's Day (December 26), when wrens were once ritually honored and hunted in symbolic processions. In European folklore, it was a bird of the hearth and home, seen as a protective spirit of small places, bringing luck and insight. Despite its modesty, the wren's bold, intricate song, often delivered from low brush or stone crevices, makes it a symbol of creative voice, resilience in obscurity, and the divine found in the everyday. As a spirit guide, the wren teaches us to use our wits, not brute force, to sing our truth no matter our size, and to understand that power is not always loud—it often whispers from the undergrowth.

Wrens are found throughout Europe, the Americas, and parts of Asia, thriving in woodlands, gardens, hedges, and marshes, especially in temperate climates. They are tiny, brown, round-bodied birds with short tails often held upright, and they are insectivorous, feeding on insects, spiders, and small invertebrates. Wrens are astonishingly vocal—the Eurasian Wren sings a song many times longer and louder than would be expected from its size, and the Carolina Wren is known for its musical, ringing notes that echo through southern U.S. woodlands. They nest in cavities or crevices and are fiercely territorial during the breeding season. As spiritual emblems, wrens remind us to stay grounded but alert, to honor our inner flame even when hidden, and to remember that great beauty often sings from the most unassuming places.

Yellowhammer
(Emberiza citrinella)

optimism and inspiration

The yellowhammer is often associated with optimism, inspiration, and sacred song. Its clear, repetitive call—often described in English folklore as “a little bit of bread and no cheese”—has made it a familiar soundtrack to rural life and a symbol of humble joy and endurance. In Celtic lore, the yellowhammer was sometimes believed to carry messages from the Otherworld, its vivid yellow plumage seen as sunlight made flesh. In Eastern European traditions, however, its association with gold and blood sometimes took a darker turn, linking the bird to omens and hidden knowledge, though this only deepened its role as a creature of mystical importance. In spiritual symbolism, the yellowhammer serves as a totem of rhythmic creativity, the power of routine infused with beauty, and finding spiritual depth in simple repetition. Its call, unwavering and sweet, reminds us that even the most familiar voice can carry divine resonance.

The yellowhammer is native to Europe and western Asia, favoring open countryside, farmlands, field edges, and scrubby hedgerows, especially in temperate climates. It is easily recognized by the bright yellow head and underparts of the male, with russet-brown streaked wings and back. Females are more subdued in tone but just as attuned to the rhythms of the land. They feed on seeds, grains, and insects, and are ground-nesters, building neat, hidden nests among grasses or low shrubs. Though their numbers have declined in some regions due to agricultural changes, the yellowhammer remains an iconic figure in British and European pastoral identity. As a spiritual guide, the yellowhammer invites us to return to our routines with renewed reverence, to sing through the ordinary with intention, and to recognize that life’s truest beauty often repeats itself like a song learned long ago—but newly sung each day.

Zebra Finch
(Taeniopygia guttata)

emotional resilience

The zebra finch is most often associated with companionship, communication, and emotional resilience. Though small in size, it represents the strength found in partnership and harmony, especially through the constant, gentle dialogue it maintains with its mate and flock. In spiritual symbolism, the zebra finch serves as a totem of connection through simplicity—a reminder that daily rituals, shared space, and quiet presence can build bonds as enduring as those forged in grand gestures. In some interpretations, its patterned plumage and vibrant cheek patches are seen as reflections of balance and creative energy, linking the zebra finch to the beauty of contrast and complementary forces. Its soft, continuous song—delicate, patterned, and conversational—echoes the idea that real strength lies not in volume, but in consistency, care, and meaningful exchange.

The zebra finch is native to Australia and parts of Indonesia, thriving in dry grasslands, scrubby plains, and open woodlands, particularly in arid and semi-arid climates. It has also adapted easily to urban environments and is among the most commonly kept finch species worldwide. The zebra finch is distinguished by the male's bold black-and-white striped throat, orange cheek patches, and coral beak, while females wear more subdued plumage. These birds feed primarily on grass seeds, insects, and greens, and are known for their cooperative breeding behaviors and strong pair bonds. They nest in shrubs, tree hollows, or man-made structures, often reusing and adapting nests with remarkable ingenuity. As spiritual guides, zebra finches invite us to nurture joy within the framework of routine, to treasure those we share our days with, and to understand that a quiet life—well-loved and well-lived—sings with its own subtle brilliance.

Zephyr Bird
(Mediterranean mythology)

embodiment of the west wind

The Zephyr Bird is not a species cataloged by science, but rather a mythical and symbolic creature, deeply rooted in the poetic and mythological traditions of the ancient Mediterranean world, particularly Greek mythology. It is believed to be the spiritual embodiment of the west wind, a living avatar of Zephyrus—the gentle god of the western breeze and harbinger of spring. While Zephyrus is known from classical myth as one of the Anemoi (wind gods), the Zephyr Bird is thought to be his companion, messenger, or even an aspect of his divine essence made flesh. Whispered about in lost verses and regional oral lore, the Zephyr Bird has come to represent transformation, invisible guidance, the return of life after stillness, and the soft voice of destiny that speaks only when we are quiet enough to listen.

According to scattered accounts in Orphic poetry, pastoral hymns, and regional Greek and Anatolian folklore, the Zephyr Bird appears only during the liminal hours—dawn and dusk, equinox and solstice, between sleep and waking. It is said to nest in the upper branches of sacred groves, particularly among laurel and olive trees, and its feathers are described as iridescent with gold, seafoam green, and soft lavender—reflecting the colors of the sky at first light. Its song is wordless but awakening, capable of stirring lost memories or dreams not yet dreamt. It never remains in one place long, always riding the westward breeze, blessing the earth with renewal, creativity, and emotional clarity. In this way, the Zephyr Bird serves not only as a herald of spring, but as a symbol of inner rebirth, arriving quietly when one is ready to let go and move forward.

Ziz
(Jewish mythology)

divine vision

Ziz represents the spiritual dominion of air and sky, the expanse of thought, revelation, and divine vision. Described in Talmudic and Midrashic texts, the Ziz is said to be so massive its feet rest on the earth while its head touches the heavens, and its wings stretch across the horizon. Unlike many destructive mythological beasts, the Ziz is seen as benevolent and protective, guarding against storms and shielding the world from chaos. In mystical interpretations, it is a keeper of secrets and cosmic order, embodying the majesty of divine wisdom and the awe-inspiring scale of the spiritual realm. Its sheer size and flight suggest the ability to transcend human perspective, offering a reminder that the greatest truths are often beyond comprehension—but not beyond reverence. As a mythic totem, the Ziz invites us to lift our gaze, embrace the mystery of what lies above, and recognize the divine orchestration in forces far greater than ourselves.

Though the Ziz has no physical counterpart, it is often compared to or inspired by real-world birds of grandeur, such as the ostrich, phoenix, or Roc, but it exists solely in the world of myth, scripture, and sacred imagination. Mentioned briefly in Psalm 50 and Psalm 80, and expanded upon in rabbinical writings, the Ziz is said to nest atop the mountains of heaven, appearing only when the world is out of balance or in need of divine intervention. As the aerial guardian in the trio of Leviathan, Behemoth, and Ziz, it completes the sacred triad of elemental dominion—sea, land, and sky—symbolizing the divine order embedded in creation. As a spiritual presence, the Ziz reminds us to respect what we cannot measure, to protect what is sacred, and to trust that sometimes the unseen wings above us are already guarding the path ahead.

ENCHANTED
TONES

www.ingramcontent.com/pod-product-compliance
Ingram Content Group UK Ltd.
Pitfield, Milton Keynes, MK11 3LW, UK
UKHW061543051026
15625UKWH00011B/586
* 9 7 8 6 5 9 8 6 1 6 3 8 0 *